Applied Time Series Analysis and Innovative Computing

Lecture Notes in Electrical Engineering

Volume 59

For other titles published in this series, go to
www.springer.com/series/7818

Sio-Iong Ao

Applied Time Series Analysis and Innovative Computing

Springer

Sio-Iong Ao
International Association of Engineers
Unit 1, 1/F, 37–39 Hung To Road
Hong Kong
publication@iaeng.org

ISBN 978-94-007-3202-5 ISBN 978-90-481-8768-3(eBook)
DOI 10.1007/978-90-481-8768-3
Springer Dordrecht Heidelberg London New York

Softcover reprint of the hardcover 1st edition 2010

Printed on acid-free paper

Springer is part of Springer Science+Business Media (www.springer.com)

To My Lovely Wife Choi, Wai-Ming

Preface

There are many reasons to analyze the time series data, for example, to understand the underlying generating mechanism better, to achieve optimal control of the system, or to obtain better forecasting of future values. Applied time series analysis consists of empirical models for analyzing time series in order to extract meaningful statistics and other properties of the time series data. With the advances in computer technology, nowadays huge amounts of time series data are stored in data warehouses. Different innovative computing techniques are needed to extract information from these datasets. Innovative computing paradigms can extract the patterns from the time series and to present it in such a way that can better our understanding of the structure, relation, and function of the subjects. The purpose of this book is to illustrate how to apply innovative computing paradigms for the applied time series analysis, with frontier application studies of the time series problems based on the recent works at the Oxford University Computing Laboratory, University of Oxford, the University of Hong Kong, and the Chinese University of Hong Kong. The monograph was drafted when the author was a postdoctoral fellow in Harvard School of Engineering and Applied Sciences, Harvard University. It provides a systematic introduction to the use of innovative computing paradigms as an investigative tool for applications in time series analysis.

In this book, innovative computing algorithms have been illustrated for solving some frontier problems in time series analysis. The book is organized as follows. In Chap. 1, it is the brief introduction to the applied time series analysis and the advances in innovative computing paradigms. The three real-word applications of innovative computing paradigms for time series problems and the contributions of these algorithms to the time series analysis are described briefly. In Chap. 2, we describe about the applied time series analysis generally. Time series analysis models including time domain models and frequency domain models are covered. In Chap. 3, we describe about the recent advances in innovative computing paradigms. Topics like computing algorithms and databases, integration of hardware, systems and networks, Internet and grid computing, and visualization, design and communication, will be covered. The advances of innovative computing for time series problems are also discussed, and an example of building of an innovative computing algorithm for some simulated time series is illustrated. In Chap. 4, we present the real-word application of innovative computing paradigms for time series problems.

The interdisciplinary innovative computing techniques are applied to understand, model and design systems for business forecasting. In Chap. 5, the second real-word application is for the analysis of the biological time series. Recurrent Elman neural networks and support vector machines have been outlined for temporal modeling of microarray continuous time series data sets. In Chap. 6, we present the last real-word application for the astronomical time series. Some innovative computing algorithms are described about how to classify the light curves of the quasars against light curves of the other stars.

In business organizations, forecasting is one of the most important activities that form the basis for strategic and operational decision. Traditionally, business time series forecasting has been dominated by linear methods. However, the linear models have serious limitation with problems of nonlinear relationships. It may be unsatisfactory to approximate the linear models for these nonlinear relationships. The importance and complexity of the business time series forecasting problem paves way for the importance of innovative computing paradigms. In Chap. 4, the interdisciplinary innovative computing techniques are applied to understand, model and design systems for business forecasting. Two business case studies are described to demonstrate the advantages of our proposed system. In global business, the interactions between different markets cause collective lead-lag behavior having special statistical properties which reflect the underlying dynamics. The internal structure of a complex system can manifest itself with correlations among its components. An innovative computing system of combining the vector autoregression and genetic algorithm with neural network is outlined how to take advantage of the lead-lag dynamics, to make the neural network forecasting process more transparent and to improve the prediction capability of the neural network.

The temporal patterns exhibited by biological time series are often complex, nonlinear or chaotic, nevertheless, the analysis are emerging as important means for many biological studies. To address the nonlinear properties of the biological time series, innovative computing algorithms are becoming important tools to study these systems. In Chap. 5, recurrent Elman neural networks and support vector machines are outlined for temporal modeling of microarray continuous time series data sets. In a single microarray experiment, the expression levels of as many as thousands of genes can be measured simultaneously. Thus, it can enable the genome-wide measurement of gene expression, and the construction of genetic network from gene expression time series with innovative computing approaches. An ensemble of the ENN and SVM models are described to better improve the transparency and robustness of the recurrent neural networks, and to further improve the prediction accuracy of the individual models. In order to provide the neural networks with explanation capabilities, a pedagogical rule extraction technique is considered for inferring the output of our proposed ensemble system. The gene regulatory network can be reconstructed satisfactorily with this hybrid innovative computing methodology. The proposed rule extraction technique provides the network inference of the ensemble. The reconstructed gene regulatory network enables us to better understand the dynamics of the underlying biological processes.

With the advances of the technologies for the sky surveys, massive amount of survey data become available. It is very helpful for the automatic and semi-automatic innovative computing methods in the classifications and detections of the astrophysical objects. In Chap. 6, the first section describes about the existing automatic and semi-automatic innovative computing methods for the comprehensive search of quasars. Quasars are interesting astrophysical objects that have been recently discovered more comprehensively from the sky surveys. Secondly, some innovative computing algorithms are described about how to classify the light curves of the quasars against light curves of the other stars.

Topics covered in the monograph include Frequency Domain, Correlation, Smoothing, Periodogram, Autoregression, ARIMA Models, Discrimination Analysis, Clustering Analysis, Factor Analysis, Dynamic Fourier Analysis, Random Coefficient Regression, Discrete Fourier Transform, Innovative Computing Algorithms, Knowledge Extraction, Large Complex Databases, Modeling and Simulations, Integration of Hardware, Systems and Networks, Grid Computing, Visualization, Design and Communication, Business Time Series Applications, Biological Time Series Applications, and Astronomical Time Series Applications. The book offers the state of art of tremendous advances in applied time series analysis and innovative computing paradigms and also serves as an excellent reference work for researchers and graduate students working on applied time series analysis and innovative computing paradigms.

The author is grateful for the supports of Dr. Vasile Palade throughout the author's research in Oxford University Computing Laboratory, University of Oxford, UK, and the supports of colleagues in Harvard throughout the author's research in Harvard School of Engineering and Applied Sciences, Harvard University, USA.

May 2009

Sio-Iong Ao, Ph.D.
International Association of Engineers
Unit 1, 1/F, 37-39 Hung To Road
Hong Kong

Contents

Chapter 1
Introduction

Abstract This book is organized as follows. In first two sections of this chapter, it is the brief introduction to the applied time series analysis and the advances in innovative computing paradigms. In the third section, we describe briefly about the three real-world applications of innovative computing paradigms for time series problems. The contributions of these algorithms to the time series analysis are also described briefly in that section and in more details in their respective chapters. In Chap. 2, we describe about the applied time series analysis generally. Time series analysis models including time domain models and frequency domain models are covered. In Chap. 3, we describe about the recent advances in innovative computing paradigms. Topics like computing algorithms and databases, integration of hardware, systems and networks, Internet and grid computing, and visualization, design and communication, will be covered. The advances of innovative computing for time series problems are also discussed, and an example of building of an innovative computing algorithm for some simulated time series is illustrated. In Chap. 4, we present the real-world application of innovative computing paradigms for time series problems. The interdisciplinary innovative computing techniques are applied to understand, model and design systems for business forecasting. In Chap. 5, the second real-world application is for the analysis of the biological time series. Recurrent Elman neural networks and support vector machines have been outlined for temporal modeling of microarray continuous time series data sets. In Chap. 6, we present the last real-world application for the astronomical time series. It is to explore if some innovative computing algorithms can automatically classify the light curves of the quasars against the very similar light curves of the other stars.

Keywords Time series analysis • Innovative computing • Frequency domain • Real-world applications

S.-I. Ao, *Applied Time Series Analysis and Innovative Computing*,
Lecture Notes in Electrical Engineering 59, DOI 10.1007/978-90-481-8768-3_1,

1.1 Applied Time Series Analysis

1.1.1 Basic Definitions

A time series can be regarded as any series of measurements taken at different times. Different from other common data analysis problems, time series data have a natural temporal ordering. Examples of time series are the daily stock prices, daily temperature, temporal gene expression values, and temporal light intensity of astronomical objects etc. Applied time series analysis consists of empirical models for analyzing time series in order to extract meaningful statistics and other properties of the time series data. Time series models usually take advantage of the fact that observations close together in time are generally more closely related than observations further apart. There are many reasons to analyze the time series data, for example, to understand the underlying generating mechanism better, to achieve optimal control of the system, or to obtain better forecasting of future values. Time series forecasting is about the employment of time series model to forecast future events based on past events. The forecasting methods have been applied in various domains, like for example, business forecasting (Ao 2003b–e, 2006, 2007b) and genomic analysis (Ao et al. 2004, Ao 2006, 2007a).

1.1.2 Basic Applied Time Series Models

Time series models have various forms and represent different stochastic processes. Different from a deterministic process, in a stochastic process, there is some indeterminacy in its future evolution described by probability distributions. Time series analysis model is usually classified as either time domain model or frequency domain model. Time domain models include the auto-correlation and cross-correlation analysis. In a time domain model, mathematical functions are usually used to study the data with respect to time. The three broad classes for modeling the variations of time series process are the autoregressive (AR) models, the integrated (I) models, and the moving average (MA) models. They all depend linearly on previous time series data points (Box and Jenkins 1976). The autoregressive fractionally integrated moving average (ARFIMA) model is the generalization of these three classes. Among the non-linear time series models, the autoregressive conditional heteroskedasticity (ARCH) models are commonly used to represent the changes of variance along time (heteroskedasticity). Here, the changes in variability are assumed to be related to recent past values of the observed series.

1.1.3 Frequency Domain Models

In a frequency domain model, the analysis of mathematical functions or signals is conducted with respect to frequency rather than time. Mathematical operators like

Fourier transform can be used to convert the time series data between the time and frequency domains, while the inverse Fourier transform can convert the frequency domain function back to a time function (Bracewell 2000). In the Fourier transform, a function is decomposed into the sum of a number of sine wave frequency components. The spectrum of these frequency components is the frequency domain representation of the function. After the transform, the counterparts of the original linear operations can often be performed more easily in the frequency domain. The Laplace transform is another popular method for the transformation from the time domain to the frequency domain, where the same inputs and outputs are functions of complex angular frequency (Davies 2002). The Laplace transform can give a simple mathematical description of an input or output to a system and may simplify the analysis of the behavior of the system. In using the Fourier or Laplace transforms, the frequency spectrum is complex and describes both the magnitude and phase of the signal as a function of frequency.

1.2 Advances in Innovative Computing Paradigms

1.2.1 Computing Algorithms and Databases

With the advances in computer technology, nowadays huge amounts of data can be stored in data warehouses. Different techniques are developed to extract information from various kinds of datasets like structured data, electronic text, and data of natural language etc. (Ao 2008a, b; Ao et al. 2005a, b; Ao et al. 2007; Ao et al. 2008). It is an active research area to extract knowledge automatically from large-scale text datasets. Information extraction is a popular field in this area. The systems focused on searching instances of pre-specified relations and can recognize entities and relations within text corpora (Banko and Etzioni 2007). Unsupervised methods have also been developed to exploit the large volumes of web pages and can extract structured knowledge from unstructured text. Knowledge discovery in databases (KDD) refers to the extraction of interesting and useful knowledge from real-world databases (Fayyad et al. 1996). A serious difficulty for the application of KDD algorithms to large complex databases is the scalability problem. Innovative computing algorithms, like neural computing, fuzzy computing, evolutionary computing, quantum computing, swarm-based computing, and DNA computing, have been developed to analyses the large-scale complex datasets and to solve new computational problems efficiently.

1.2.2 Integration of Hardware, Systems and Networks

Innovative experimental hardware system includes quantum computer, chemical computer, DNA computing, optical computer, and spintronics based computing etc. In a chemical computer, the computations are performed by naturally occurring chemical reactions. An optical computer is a devise that utilize the photons in visible light or infrared beams to perform digital computation. Spintronics technology utilizes the intrinsic spin of the electron, its associated magnetic moment, and its fundamental electronic charge, in solid-state devices. The data-acquisition devices act as the interfaces between the signal and a computer. Data-acquisition devices convert physical parameters to electrical signals that can be manipulated by a computer. The visualization is an interdisciplinary study based on research in human–computer interaction, computer science, graphics, visual design and psychology, and becomes an important component in scientific research (Bederson and Shneiderman 2003). Visual displays can help the users to explore the information in visual form like 3D outputs. The interaction devices provide a way for computer users to accomplish their tasks. The graphics processing unit (GPU) in the form of graphics card is a common modern PC co-processor. Not only are the GPU very efficient at manipulating computer graphics, they are an attractive platform for general purpose computation. Network interoperability refers to the continuous ability to transfer data between interconnected networks.

1.2.3 Internet, Web and Grid Computing

It is a challenging problem to study communication and data sharing in highly dynamic distributed environment. In distributed computation, the system consists of autonomous computers that communicate through a computer network. Traditional communication methods like hardware videoconference can only serve the communication between a few persons. Collaboration often presents unforeseen social and organizational challenges (Olson et al. 2008). Grid computing refers to the combination of computer resources from multiple administrative domains applied to a common computational task. A difficulty in grip computing is that the designer must have measures to prevent malfunction in any participating computer from producing erroneous results. The pooling of the remote computer resources can make a collection of computer resources behave like a single pooled resources. Pooling techniques have already been used commercially in the content distribution networks of Akamai and Google etc.

1.2.4 Visualization, Design and Communication

For scientific projects, it is also a challenging task to have effective communication within the scientific team. Knowledge visualization and interactivity is an essential issue with regard to organizational communication. Visualization techniques and tools like ParaView, VisIT, POVRay, volume renderin, particle rendering, are used in real-time analysis and filtering of the data. Good graphical displays should include characteristics like inducing the viewer to think about the substance. In commercial applications like flight traffic control, good three-dimensional flight display can increase the efficiency of remote sensing flight operations. Displaying of complex information is also useful in scientific works. For some scientific researches, it is important not to be restricted within the studies of two-dimensional components. Scientists are now using more interpretive graphic art than ever to convey their ideas and results. New visual techniques are developed to better understand and communicate the experimental and computational results. For many scientific projects, it is an important step to analyze large and distributed image datasets in order to help understand the characteristics of the systems concerned. For real-time image processing, specialized data paths and pipelining are typically required. Widely used approaches for representation of uncertainties include interval mathematics, classical set theory, fuzzy theory, and probabilistic analysis (Klir 1994). Different modeling techniques have been used to solve the uncertainty in real-world problems.

1.3 Real-World Applications: Innovative Computing Paradigms for Time Series Problems

1.3.1 Developing Innovative Computing Algorithms for Business Time Series

Business forecasting is a very difficult task, because the processes can behave more like a random walk process and may be time varying. The linear models have serious limitation with problems of nonlinear relationships. The interdisciplinary innovative computing techniques can be applied to understand, model and design systems for business forecasting. The neural network (NN) is well-known for its capability of pattern recognition and has been playing a more and more active role in the forecasting field. The adopting of neural networks for time series analysis began in the mid-1980s, and then neural networks emerged as an important tool for business forecasting. Evolutionary computing algorithms, like genetic algorithm (GA), have also proved its applicability for time series forecasting in a number of studies. Due to the complex nature of the business time series, it is sometimes difficult to apply a single intelligent technique to model the characteristics of the time series. The vector autoregression (VAR) techniques (Enders 1995; Greene 2000) are used to understand their interactions among different variables. The VAR model is a useful modeling method for business forecasting.

Hybrid intelligent computing models consisting of different innovative components are developed with the objective to combine their respective advantages. Modular hybrid systems refer to those systems that are modular in nature. Our works of the hybrid vector autoregression, neural network and genetic algorithm (VAR-NN-GA) framework can supplement its separate stand-alone components (Ao, DOI: 10.1007/s00500-010-0580-4). The objective of our framework is to develop a system that can take over the task of the model selection process. The vector autoregression (VAR) models are employed to test the interdependence of the time series variables. The neural network was an integral part of our knowledge discovery and forecasting process and was central to the proposed methodology. The goal of the GA is to let the strategy with higher payoff to remain while the ones with lower payoff more likely to disappear.

1.3.2 Developing Innovative Computing Algorithms for Biological Time Series

The temporal patterns exhibited by biological time series are often complex, nonlinear or chaotic, nevertheless, the analysis are emerging as important means for many biological studies. To address the nonlinear properties of the biological time series, nonlinear methods become the important tools for the studies. Among the high-throughput methods of gene expression, the microarray has been the most widely used one. The microarray technology can also enable us to have the gene expression values at different time points of a cell cycle. The frequencies of the time points can have ranges from several minutes to several hours. Various methods like self-organizing maps (Nikkilä 2002), *k*-nearest neighbor (Acta 2001) and hidden Markov models (Ji et al. 2003) have been employed for the microarray analysis. These studies mainly focus on the clustering and the measurement of the similarity among the different expressions.

In our works, recurrent Elman neural networks (ENN) and support vector machines (SVM) have been outlined for temporal modeling of microarray continuous time series data sets. The Elman network is a recurrent neural network of simple architecture which can be trained with the backpropagation (BP) algorithm. Recurrent neural networks have feedback links and incorporate temporal aspects better than feedforward neural networks (Kodogiannis and Lolis 2002). Support vector machines have many mathematical features attractive for gene expression analysis (Brown et al. 2000). Ensemble methods have been developed for improving the predictive performance of a given individual statistical learning algorithm. Previous studies have shown that ensemble methods can often provide more accurate prediction than any of the individual classifiers in the ensemble (Opitz and Maclin 1999). An ensemble of the ENN and SVM models are described for how to further improve the prediction accuracy of the individual models. Systems Biology studies on gene interactions have gained its prominence for reconstructing gene regulatory networks using the microarray technology (Cho et al. 2007). In order to provide the neural networks with explanation capabilities, a pedagogical rule extraction technique is described in our works for inferring the output of our proposed ensemble system.

1.3.3 Developing Innovative Computing Algorithms for Astronomical Time Series

The astronomical time series include temporal data of astronomical objects like interacting binaries, planet searches, pulsars, active galactic nuclei, quasars, and gravitational lensing etc. The advances of the technologies for the sky surveys significantly increase the quantity, quality and variety of the time series data of the astronomical events (Maoz et al. 1997). Traditionally, the classification is conducted by the human expert's unaided eye-brain. For surveys of millions of objects, it may not be possible to detect the desired objects by expert inspection alone. It would be very helpful for the automatic and semi-automatic methods in the classifications/detections of the astrophysical objects. Nowadays, a variety of classification methods have been proposed for the astronomical time series. The goal is to create a system that can classify the astronomical time series automatically with the performance as well as human experts. Another common task of the astronomical time series is to find the patterns in the temporal data. Clustering methods can efficiently find patterns that do not rely on a regular periodicity.

Quasars are interesting astrophysical objects that have been recently discovered more comprehensively from the sky surveys. Quasars, quasi-stellar radio sources, are powerfully energetic and distant galaxy with an active galactic nucleus. The studies of quasars may help us better understand the early stage of the universe. The quasars can also serve as suitable reference frames against which to measure the proper motion of other astrophysical objects (Geha et al. 2003). Thus, innovative approaches have been applied for the investigation of the quasars light variation, which is shown to be helpful for the automatic quasar selection. In our exploratory works, it is to see if some machine learning algorithms can automatically classify the light curves of the quasars against the very similar light curves of the other stars. For example in MACHO sky survey, the light curves of the Be stars are so similar with the quasar light curves that the previous algorithms and even the manual examination by experts cannot tell the difference between the light. Experimental results will also be shown for the applied exploratory innovative algorithms.

Chapter 2
Applied Time Series Analysis

Abstract There are many reasons to analyze the time series data, for example, to understand the underlying generating mechanism better, to achieve optimal control of the system, or to obtain better forecasting of future values. Applied time series analysis consists of empirical models for analyzing time series in order to extract meaningful statistics and other properties of the time series data. Time series models have various forms and represent different stochastic processes. Time series analysis model is usually classified as either time domain model or frequency domain model. Time domain models include the auto-correlation and cross-correlation analysis. In a time domain model, mathematical functions are usually used to study the data with respect to time. The three broad classes for modeling the variations of time series process are the autoregressive models, the integrated models, and the moving average models. The autoregressive integrated moving average models are the general class of these models for forecasting a time series that can be stationarized by transformations such as differencing. In a frequency domain model, the analysis of mathematical functions or signals is conducted with respect to frequency rather than time. Mathematical models can be used to convert the time series data between the time and frequency domains. The parameters and features in the frequency domain can be used as inputs for the mathematical models like discrimination analysis and improved results can be obtained.

Keywords Time series analysis • Basic characteristics • Estimation of correlation • Autoregression • Mathematical models

S.-I. Ao, *Applied Time Series Analysis and Innovative Computing*,
Lecture Notes in Electrical Engineering 59, DOI 10.1007/978-90-481-8768-3_2,

2.1 Basic Characteristics of Time Series

2.1.1 Estimation of Correlation

2.1.1.1 Auto-Correlation Analysis

The autocorrelation of a time series measures the correlation between values of the time series at different points in time, as a function of the two times. It can be used to find repeating patterns, or to identify the missing fundamental frequency in a signal implied by its harmonic frequencies. It is a popular tool for analyzing time domain signals. Mathematically, the autocorrelation between any two time s and t is defined by

$$R(s,t) = \frac{E\left[(X_t - u_t)(X_s - u_s)\right]}{\sigma_t \sigma_s}$$

where X_i is the value of a given time series X at time i, u_i and σ_i^2 are the mean and variance for all time i. If the mean and variance of the time series are time-independent, the autocorrelation can be simplified as

$$R(\tau) = \frac{E\left[(X_t - \mu)(X_{t+\tau} - \mu)\right]}{\sigma^2}$$

where the time-lag $\tau = t - s$. For a discrete process of length $n\{X_1, X_2, ..., X_n\}$ with known mean and variance, an unbiased estimate of the autocorrelation is

$$\hat{R}(k) = \frac{1}{(n-k)\sigma^2} \sum_{t=1}^{n-k} [X_t - \mu][X_{t+k} - \mu]$$

for any positive integer $k < n$. If the mean and variance are replaced by the standard formulae for sample mean and sample variance, then it is a biased estimate (Priestley 1982). A fundamental property of the autocorrelation is symmetry, i.e. $R(i) = R(-i)$. Another property is that the continuous autocorrelation function reaches its peak at the origin. It can also be shown that the autocorrelation of a periodic function is also periodic with the same period.

2.1.1.2 Cross-Correlation Analysis

Cross-correlation is a measure of similarity of two time series as a function of a time-lag applied to one of them. It can be applied to search a long duration signal for a shorter, know feature, and has applications in areas like pattern recognition and single particle analysis, etc. Mathematically, the cross-correlation is defined as

$$(f \bullet g)(t) = \int_{-\infty}^{\infty} f^*(\tau) g(t+\tau) d\tau$$

for continuous time series functions, *f* and *g*, where f^* is the complex conjugate of *f*. For discrete functions, the cross-correlation is defined as

$$(f \bullet g)[n] = \sum_{m=-\infty}^{\infty} f^*[m]\, g[n+m]$$

For two independent random variables *X* and *Y* with probability distributions *f* and *g*, the probability distribution of the difference *Y–X* is given by the cross-correlation $f \bullet g$. A property of the cross-correlation of function *f(t)* and *g(t)* is $(f \bullet g) \bullet (f \bullet g) = (f \bullet f) \bullet (g \bullet g)$. The cross-correlation of a time series with itself is the autocorrelation, and there will always be a peak at a lag of zero. In template matching, the normalized correlation can be defined as (Lewis 1995):

$$\gamma(u,v) = \frac{\sum_{x,y}\left[f(x,y) - \bar{f}_{u,v}\right]\left[t(x-u,y-v) - \bar{t}\right]}{\left\{\sum_{x,y}\left[f(x,y) - \bar{f}_{u,v}\right]^2 \sum_{x,y}\left[t(x-u,y-v) - \bar{t}\right]^2\right\}^{0.5}}$$

where $\bar{t}$ is the mean of the feature and $\bar{f}_{u,v}$ is mean of *f*(*x*,*y*) in the region under the feature.

2.1.1.3 Autocorrelation Functions

The autocorrelation function can identify the time interval over which a correlation in the noise exists. As said in the first section, autocorrelation is the correlation of a time series set with itself, offset by *n*-values. The lag *k* autocorrelation function is the set of autocorrelations with offsets 1, 2, 3, 4,…, *k*. Mathematically, the lag *k* autocorrelation function is defined as

$$\tau_k = \frac{\sum_{i=1}^{N-k}\left(f(i) - \bar{f}\right)\left(f(i+k) - \bar{f}\right)}{\sum_{i=1}^{N}\left(f(i) - \bar{f}\right)^2}$$

where the observations are equi-spaced.

The autocorrelation function can be used to check the non-whiteness of data or residuals, to detect periodic components in data, and to identify the dominant power law noise type. The autocorrelation function of a periodic signal is itself a periodic signal, with a period the same as that of the original signal. Short signals can have short autocorrelation functions as well, while the autocorrelation functions of random noise is a single sharp spike at shift zero. It is interesting to note that the autocorrelation function and the power spectrum are related by the Fourier transform (Percival and Walden 1993). When the Fourier transform is applied to the autocorrelation function, the power spectrum can be obtained.

2.1.2 Stationary Time Series

A time series is said to be stationary if its statistical properties do no depend on time. A time series may be stationary with respect to one characteristic, while not stationary with respect to another. A time series is said to be weakly stationary if its mean, variance and autocovariance do not grow over time. For example, there can be a time series where its mean does not depend on time while its variance depends on time. As an example, white noise is stationary. A time series is said to be a white noise process if each value in the time series have zero-mean, constant conditional variance and is uncorrelated with all other realizations. Another example is the autoregressive moving average model, which is discrete-time stationary process with continuous sample space. Many business time series are far from stationary. Nevertheless, stationary time series can provide us with meaningful sample statistics such as means, variances, and correlations with other variables. Also many statistical forecasting methods are based on the assumption that the time series are approximately stationary. Mathematical transformations like deflation, seasonal adjustment, and de-trending can be applied to try to stationarize the time series.

2.1.3 Smoothing of the Time Series

The irregularities and random fluctuations in time series data often obscure the efforts to understand the process being observed. Smoothing techniques are employed to address these problems and to make long term fluctuations in the time series stand out more clearly. The moving average smoothing is the most common type of smoothing techniques. The technique can reduce random fluctuations and make adjustments for seasonal or cyclical components of a time series. Mathematically, the moving average smoothed statistic s_t is the mean of the last k observations:

$$s_t = \frac{1}{k}\sum_{n=0}^{k-1} x_{t-n}$$

where the choice of an integer $k > 1$ is arbitrary. A slightly modified version of the simple moving average technique is the weight moving average, which is defined as

$$s_t = \sum_{n=1}^{k} w_n x_{t+1-n}$$

where the set of weighting factors satisfy the constraint $\sum_{n=1}^{k} w_n = 1$.

Another popular smoothing technique is the exponential smoothing. It can reduce the effects of random fluctuations in the time series and in many cases is an effective tool in time series forecasting. In moving average smoothing, all the past observations are weighted equally, whereas in exponential decreasing weights are assigned as the observations get older. Mathematically, the exponential smoothing can be computed with the algorithm:

$$s_t = \alpha x_t + (1-\alpha)s_{t-1}$$

where $s_0 = x_0$, and α is the smoothing factor with the constraint $0 < \alpha < 1$.

2.1.4 Periodogram Analysis

The periodogram is an estimate of the spectral density of a signal (Schuster 1898). Mathematically, for an ordered set of N random variables $X_1, X_2, \ldots, X_N$, the periodogram can be computed as followed (Herbst 1963):

$$I_x(\lambda) = 2\left|J_x(\lambda)\right|^2 \quad \left(0 \le \lambda \le \tfrac{1}{2}\right)$$

where $I_x(\lambda)$ is computed at a finite set of λ-values, usually

$$\lambda = s/N \left(s = 0,1,2,\ldots,\left[\tfrac{1}{2}N\right]\right),$$

$$J_x(\lambda) = N^{-1/2}\sum_{t=1}^{N} X_t z^{t\lambda} = N^{-1/2}\sum_{t=1}^{N} X_t \cos 2\pi t\lambda + iN^{-1/2}\sum_{t=1}^{N} X_t \sin 2\pi t\lambda$$

with $z^{\lambda} = e^{2\pi i\lambda}$ for real finite λ, and, for real random variables U, V,

$$|U + iV|^2 = U^2 + V^2.$$

In raw periodogram, it may not be a good spectral estimate because the variance at a given frequency does not decreased as the number of samples used in the computation increases. The smoothing of the periodogram can be used to address this variance problem. When smoothing is applied to reduce the effect of measurement noise, the smoothed version of the periodogram is also called the spectral plot. Another technique for this problem is the method of average periodograms (Engelberg 2008). In practice, the periodogram is often computed from a finite-length time series using the fast Fourier transform (FFT).

2.2 Autoregression and ARIMA Models

2.2.1 *Time Series Regression*

Linear regression is useful for exploring the relationship of an independent variable to a dependent variable when the relationship is linear (O'Sullivan and Rassel 1999). However, there are problems when there is the autocorrelation in a time series. In the time series, the values at one time may be influenced by values at a previous time. This happens when the values of the dependent variable over time are not randomly distributed. Another problem is that linear regression fails to capture seasonal, cyclical, and counter-cyclical trends in time series. Regression options like transforming variables with logarithms, differences, inverses, lags, and de-trending can be applied for the regression analysis. The form in the regression relies ultimately upon the time series process of the variables concerned. The theory can serve as a guide to the variable selection, while the regression results may lead to ways to refine the underlying theory.

Time series regression models are very suitable for evaluating short-term effects of time-varying exposures. Nevertheless, there are problems that may occur in the regression models, like multicollinearity, heteroscedasticy, and autocorrelation (Ostrom 1990). The Durbin-Watson statistic can be employed to test for the presence of first-order autocorrelation in the residual of a regression model. Treatments for autocorrelated error terms include the finding of an important omitted variable, transforming of the variables based upon generalized least squares, and introducing of the time as a dependent variable on the regression model.

A popular statistical package for time series analysis is the RATS, an abbreviation of Regression Analysis of Time Series (http://www.estima.com). It is relatively inexpensive as compared with other similar commercial packages. The package includes the major procedures in econometrics and time series analysis, like linear regression, regressions with heteroscedasticity and serial-correlation correction, non-linear least squares, generalized method of moments, maximum likelihood estimation, ARIMA, spectral analysis, Kalman filter, logistic regression, ARCH and GARCH models, and vector autoregressions. Another popular statistical computing package is the R software environment. Its source code is freely available under the GNU General Public License, and pre-compiled binary versions are available for operating systems (http://www.r-project.org). R supports statistical computing like linear and nonlinear modeling, classical statistical tests, time series analysis, classification and clustering, etc.

2.2.2 *Autoregressive Moving Average Models*

The autoregressive moving average model consists of the autoregressive (AR) and the moving average (MA) parts, and is referred as the ARMA(p,q) model, where p is the order of the autoregressive part and q is the order of the moving average part (Box et al. 1994). Mathematically, the AR(p) model can be expressed as

$$X_t = c + \sum_{i=1}^{p} \varphi_i X_{t-i} + \varepsilon_t$$

where c is a constant, $\varphi_1, \ldots, \varphi_p$ are the parameters of the model, and ε_t is the white noise. In the AR(1) model, the model is not stationary for $|\varphi_1| \geq 1$. Mathematically, the moving average MA(q) model can be expressed:

$$X_t = \mu + \varepsilon_t + \sum_{i=1}^{q} \theta_i \varepsilon_{t-i}$$

where μ is the expectation of X_t, $\theta_1, \ldots, \theta_q$ are the parameters of the model, and the $\varepsilon_t, \varepsilon_{t-1}, \ldots$ are the white noise error terms. Combining the AR(p) and MA(q) models, we can have the mathematical expression of the ARMA(p, q) model:

$$X_t = c + \varepsilon_t + \sum_{i=1}^{p} \varphi_i X_{t-i} + \sum_{i=1}^{q} \theta_i \varepsilon_{t-i}$$

In the above models, the error terms are assumed to be independent identically-distributed random variables (i.i.d.) sampled from a normal distribution with zero mean: $\varepsilon_t \sim N(0, \sigma^2)$, where σ^2 represents the variance. These ARMA models are univariate models, and extensions for the multivariable cases are the vector autoregression (VAR), which we will talk about in later chapter.

2.2.3 *Building ARIMA Models*

ARIMA (autoregressive integrated moving average) models are the general class of models for forecasting a time series that can be stationarized by transformations such as differencing. The first step in the ARIMA procedure is to ensure the series is series is stationary (Kosuke et al. 2008). The additional differences are applied until the diagnostics indicate stationarity. This is to derive the stationary time series. The lagged values and model errors of the stationary time series are used to model the time series. The fine-tuning of the model is accomplished by adding lags of the differenced series and/or lags of the forecast errors to the prediction, as necessary so that the autocorrelation from the forecasting errors can be removed as much as possible. ARIMA utilizes a state space representation of the ARIMA model to assemble the likelihood and then utilizes the maximum likelihood to estimate the parameters of the model (Brockwell and Davis 1991).

ARIMA models can be expressed as ARIMA(p,d,q), where p is the number of autoregressive terms, d is the number of non-seasonal differences, and q is the number of lagged forecast error in the prediction equation. ARIMA(0,1,0) is the random walk process, while ARIMA(1,1,0) is the differenced first-order autoregressive model. ARIMA(0,1,1) is the simple exponential smoothing model, while ARIMA(0,2,1) or ARIMA(0,2,2) is the linear exponential smoothing model. Mathematically, the ARIMA(*p*,*d*,*q*) model of the time series $\{x_1, x_2, \ldots\}$ is defined as

$$\Phi_p(B)\Delta^d x_t = \Theta_q(\mathbf{B})\varepsilon_t$$

where B is the backward shift operator, $Bx_y = x_{y-1}$, $\Delta = 1 - B$ is the backward difference, and Φ_p and Θ_q are polynomials of order *p* and *q* respectively. It can be observed that the model consists of three different parts, i.e. an autoregressive part AR(*p*) $\Phi_p = 1 - \phi_1 B - \phi_2 B^2 - \ldots - \phi_p B^p$, an integrating part $I(d) = \Delta^{-d}$, and a moving average MA(*q*) part $\Theta_q = 1 - \theta_1 B - \theta_2 B^2 - \ldots - \theta_q B^q$.

2.2.4 *Forecasting and Evaluation*

The above models can help us to better understand the time series data, and to forecast future points in the time series. As these models with lagged values are often used for forecasting, researchers have looked for measures that can produce more accurate results for out-of-sample forecasting (Greene 2000). Instead of the adjusted R^2 (sum of squares of the model errors), measures like Akaike information criterion (AIC), and Schwartz Bayesian criterion (SBC) are utilized, with

$$\text{AIC} = T\ln\left(\text{residual sum of squares}\right) + 2n$$

$$\text{SBC} = T\ln\left(\text{residual sum of squares}\right) + n\ln(T)$$

where n is the number of parameters estimated (p + q + possible constant term), and T is the number of usable observations. The value of n yielding the minimum AIC value specifies the best model, and AIC tends to be superior in smaller samples where the relative number of parameters is large (McQuarrie and Tsai 1998). The idea behind these measures is that incorporating additional coefficients will necessarily increase fit at a cost of reducing degrees of freedom (Enders 1995). The objective is to have parsimonious model that fits the data well without incorporating any needless coefficients. The parsimonious models can generally produce better forecasts than over-parameterized models.

2.2.5 *Causality of the Time Series*

The concept of the causality of the time series was introduced by Wiener (1956) and Granger (1969) and become a popular notion for studying the dynamic relationships between time series. A time series variable *Y* is said to cause *X* in the sense of Granger if the observation of Y up to time t can help to predict $X(t + 1)$. Recent researches noted that it is possible that Y does not cause X in the above sense, but can still help the prediction of X several periods ahead (Sims 1980; Giles 2002). This may be due to the fact that Y may be able to help the prediction of another variable one period ahead, and subsequently that variable has an effect on X at a later period. It is helpful to study these indirect effects in order to better analyze the relationships between time series.

A direct way to determine Granger causality is to use a standard F-test to test whether the lags of one variable y enter into the equation for another variable x. If all the coefficients for the lags of y are equal to zero in that equation, it can be shown that y does not Granger cause x (Hamilton 1994). Mathematically, this can be expressed as bivariate vector autoregression, with

$$x_t = c_1 + \sum_{i=1}^{p} \alpha_i x_{t-i} + \sum_{i=1}^{p} \beta_i y_{t-i} + u_t$$

with lag length p. An F-test of the following null hypothesis is set up

$$H_0 : \beta_1 = \beta_2 = \ldots = \beta_p = 0$$

$$x_t = c_t + \sum_{i=1}^{p} \gamma_i x_{t-i} + e_t$$

A test statistic S_1 can be obtained as followed:

$$S_1 = \frac{(RSS_0 - RSS_1)/p}{RSS_1/(T-2p-1)} \sim F_{p,T-2p-1}$$

where $RSS_1 = \sum_{t=1}^{T} u_t^2$ and $RSS_0 = \sum_{t=1}^{T} e_t^2$. If the test statistic is greater than the specific critical value, the null that y does not Granger-cause x is rejected.

2.3 Mathematical Models in the Frequency Domain

2.3.1 *Introduction*

In the frequency domain, the analysis of mathematical functions or signals is conducted with respect to frequency rather than time. Mathematical models can be used to convert the time series data between the time and frequency domains. A time series can be decomposed into cyclic components, which can be described by their periods and frequencies. The frequency of a cycle is the number of occurrences in a fixed time unit. The analysis in the frequency domain can detect these cycles and compute their corresponding frequencies. The Fourier transform is a popular transforming tool to compute the frequency domain representation of a time series. It can generate a description of the distribution of the energy (the amplitude) of the time series as a function of frequency. The plot of this amplitude (y-axis) against the frequency (x-axis) is called the spectrum. In practice, the Fast Fourier Transform (FFT) is often used as a quick way of performing this transform. The parameters and features in the frequency domain can be used as inputs for the mathematical models like discrimination analysis and improved results can be obtained (Voss et al. 1996; Jing and Zhang 2004).

2.3.2 *Discrimination Analysis*

Discrimination analysis is a method for classifying a set of observations into predefined classes, based on a set of input variables know as predictors. It is very useful for the detection of variables that can discriminate between different groups efficiently, and for the classification of new cases into different groups (Hill and Lewicki 2007). The model is built from a training set of observations with known classes. Computationally, discrimination analysis is very similar with the analysis of variance (ANOVA). Its underlying idea is to decide whether groups differ with regard to the mean of a variable, and then to utilize that variable for the prediction of group membership of new cases. As a simple example, suppose that there is a random sample of 100 adults and 100 children. On the average, adults are heavier than children. Thus, the variable weight can be utilized to discriminate between adults and children with a better than chance probability. In practice, several variables may be considered in a single study to check which variables contribute to the distribution between groups. Multivariable F tests can be used to test whether there are any significant differences with regard to the variables between groups.

The objective of discrimination analysis is to find the variables that discriminate between groups. There are several popular models to implement the discrimination analysis, including the forward stepwise analysis, backward stepwise analysis, and F to enter, F to remove technique, etc. In forward stepwise analysis, at each step, all the available variables are checked to see which one contributes most to the discrimination between groups. Then that variable will be included in the model, and the testing will proceed to the next step with all the other still available variables. In the backward stepwise analysis, firstly all variables are included in the model. Then, at each step, the variable that contributes least to the predicting of group membership will be eliminated. The procedure will stop when the contribution of each remaining variable to the membership prediction is higher than the predefined threshold value. In the F to enter, F to remove strategy, it is similar with the stepwise multiple regression procedure, and the stepwise procedure follows the F to enter and F to remove values.

The discrimination analysis can be applied in bankruptcy prediction, face recognition, marketing research, and product management, etc. (McLachlan 2004). The analysis can also be applied to the input variables in the frequency domain. For example, Jing and Zhang (2004) proposed a new face and palmprint recognition approach based on the discrete cosine transform (DCT) and discrimination analysis. A two-dimensional separability judgment was used to select the DCT frequency bands with favorable linear separability. Then, the linear discriminative features were extracted from these selected bands for the classification process. The procedure can result in better classification performance, significantly improve the recognition rates and effectively reduce the dimension of feature space.

2.3.3 Clustering Analysis

The goal of the clustering algorithms is to figure out the underlying similarities among a set of feature vectors $\boldsymbol{x}$, and to cluster similar vectors together (Theodoridis and Koutroumbas 2003). The clustering process can also be called unsupervised pattern recognition. This is different from the supervised pattern recognition, in which a set of training data is available, and which the classification algorithms can exploit this known information in advance. The clustering algorithms have many different applications in social sciences, engineering and medical science.

The clustering process can be viewed as a combinatorial problem of putting the data points into optimal clusters. However, it is NP-hard to enumerate all such possibilities of clustering. Let $S(N, m)$ be the number of all possible clustering of N vectors into m groups. We can easily see that $S(N, 1) = 1$, $S(N, N) = 1$, and $S(N, m) = 0$, for $m > N$. It satisfies the following recursive relationship (Spath 1980):

$$S(N,m) = mS(N-1, m) + S(N-1, m-1)$$

Its solution is found to be the Stirling numbers of the second kind (for details, see (Liu 1968), etc.):

$$S(N,m) = \frac{1}{m!}\sum_{i=0}^{m}(-1)^{m-i}\binom{m}{i}i^N$$

We can see that the solutions for this problem explore exponentially with (Spath 1980): $S(15,3) = 2.38 \times 10^6$, $S(20,4) = 4.52 \times 10^{10}$, and $S(100,5)$ is of order 10^{68}. It is impractical to enumerate all possible clusters for all possible values of m.

The clustering algorithms can also allow us to infer some hypothesis concerning the nature of the data. It can be a tool for suggesting hypothesis (Theodoridis and Koutroumbas 2003). These hypotheses can be verified by using other data sets as validation sets. Another use is on the prediction that bases on groups. The algorithms can provide us with clusters that are characterized by the similarity of vectors within each cluster. When a new data set or pattern is available, we can assign it to the known cluster by comparing its characters with each cluster's characters. Clustering is important for the data reduction purpose too. There are many times that the amount of the data is very large and it is expensive to process all the data. Cluster analysis can be employed for grouping the data into a number of clusters, and then we can process each cluster as a single element.

In view of the computational difficulty, different clustering algorithms have been developed so that only a small number of the different possible combinations of the clusters will be considered. There are four main types of clustering algorithms: sequential algorithms, hierarchical clustering algorithms, clustering algorithms with cost function optimization, and others (like branch and bound algorithms, and genetic clustering algorithms). Different clustering algorithms usually produce different clustering results. It may depend on the problem to decide which type of clustering algorithms is employed.

In designing the clustering algorithm for solving a problem, a major issue is on how to define the similarity between two feature vectors. It is important to choose an appropriate measure for this task. Then, it is also important to choose an appropriate algorithmic scheme that clusters the vectors, basing on the selected similarity measure. Generally speaking, different results can be obtained with different algorithmic schemes. Expert opinions are often needed for the interpretation of the results and for choosing a suitable scheme. In our project, experts from the Genome Research Centre have provided us expert opinions on this subject.

The concept of distance measure is important for the clustering process, which need this measurement of the mathematical distance between individual observations, and groups of observations (Finch 2005). Distance in this context can be in the Euclidean sense, or some other comparable conceptualization like Manhattan distance, Hamming distance, etc. This will affect the shape of the clusters, as some objects may be close to one another with one distance while further away with another distance. A primary assumption underlying these distance measures is that the variables are continuous in nature. Finch discussed about the distance measures in Cluster analysis with dichotomous data. The definition of a distance measure has an important role in the evaluation of clustering algorithms of gene expression profiles. Ido et al. (2007) compared different clustering solutions when using the Mutual Information (MI) measure, Euclidean distance and Pearson correlation coefficient.

Clustering process can be grouped as hierarchical or partition clustering. Hierarchical clustering find successive clusters based on previously established clusters. Partition clustering establishes all clusters at once. Hierarchical clustering can be further divided into two basic groups: agglomerative and divisive clustering. Agglomerative clustering is of the bottom-up approach, that is, starts with each object as a separate cluster and then merge the objects into successively larger clusters. On the other hand, the divisive clustering is of the top-down approach. It starts with the whole set and then divide it into successively smaller clusters. Co-clustering is a clustering technique that not only clusters the objects themselves, but also their features as well. Another difference among the clustering algorithms is to look at whether the clustering process uses symmetric or asymmetric distances.

For example, Euclidean distances are symmetric, which mean distance from object H to K is the same as the distance from K to H. In applications like sequence-alignment methods, asymmetric distances have been used (for example, Prinzie and Van den Poel 2006).

The partition clustering algorithms usually determine all the clusters at once. K-means clustering, fuzzy c-means clustering and derivatives are popular partition algorithms. In the K-means algorithm, initially there are k randomly generated clusters. The average of all the points in a cluster is assigned as the center. Then, the data point is assigned to the nearest cluster center, and the new cluster centers are computed. The above two steps are repeat until the convergence criterion is met. In fuzzy clustering, each data point belongs to a cluster to a certain extent, rather than to one certain cluster completely. Besides this difference, the fuzzy c-means algorithm is similar with the k-means algorithm. Like hierarchical clustering, the partition clustering algorithms are also very popular for the genomic analysis, with microarray data sets, etc.

In spectral clustering, the dimensionality reduction for clustering in lower dimensions is performed with the spectrum of the similarity matrix of the data. A popular spectral clustering is the Shi-Malik algorithm, which is widely used for image segmentation. Liu et al. (2008) applied the spectral clustering to the analysis of correlation mutations in HIV-1 protease. The spectral clustering of the resulting covariance matrices disclosed two distinctive clusters of correlated residues. Oliveira and Seok (2008) proposed a multilevel spectral algorithm which can identify protein complexes more accurately with less computational time. Winter et al. (2004) studied the problem of blind source separation (BSS) with complex valued sample data in the frequency domain. High quality separation of speech sources is important for the further processing like speech recognition, and BSS is about the finding of the unknown underlying mixing process. Winter et al. proposed an algorithm based on hierarchical clustering to estimate the mixing matrix. The algorithm directly works on the complex valued sample data in the frequency domain and shows good convergence. Accurate estimations of the mixing matrix and very low musical tone noise were achieved.

2.3.4 Principal Components and Factor Analysis

Among the tools of the dimension reduction and transformation, the principal component analysis (PCA) is a popular tool for many researchers. Its basic idea is to find the directions in the multidimensional vector space that contribute most to the variability of the data. The principal component analysis was applied to reduce the dimensionality of the gene expression data in studies (Taylor et al. 2002; Yeung and Ruzzo. 2001, etc.). The focuses are on the effective dimensional reduction by the PCA, the analysis of the compressed space and the assistance of the PCA for the classification and the clustering. Khan et al. (2001) applied the PCA and neural network for the classification of cancers using gene expression profiling.

Its basic idea is to find the directions in the multidimensional vector space that contribute most to the variability of the data. The representation of data by the PCA consists of projecting the data onto the k-dimensional subspace according to

$$x' = F(x) = A^t x$$

where x' is the vectors in the projected space, A^t is the transformation matrix which is formed by the k largest eigenvectors of the data matrix, x is the input data matrix. Let $\{x_1, x_2, \ldots, x_n\}$ be the n samples of the input matrix x. The principal components and the transformation matrix can be obtained by minimizing the following sum of squared error:

$$J_k(a, x') = \sum_{h=1}^{n} \left\| \left(m + \sum_{i=1}^{k} a_{hi} x_i' \right) - x_h \right\|^2$$

where $m$is the sample mean, x_i' the ith largest eigenvector of the co-variance matrix, and a_{hi} the projection of x_h to x_i'.

Principal component analysis can be applied with data in the frequency domain. As an example, it is used as a general exploratory tool which leads to specific generator models in quantitative electroencephalography (qEEG). A great part of qEEE is about the computing and interpreting of EEG spectra in the frequency domain (Valdes et al. 1992). Valdes et al. proposed a new method for estimating sources in the frequency domain which fitted dipoles to the whole crosspectrum. The PCA was used for the exploratory data analysis, and the spherical harmonic functions are used to explain the recurring pattern of maps characteristic of the spatial PCA of qEEG data. Wu et al. (1998) applied the frequency spectrum PCA for modeling the sound frequency distribution features. The frequency spectra of the found were treated as a vector in a high-dimensional frequency feature space. The PCA was used to compute the variance distribution for the frequency vectors, with the largest eigenvalues accounting for the most variance within the data set. The proposed method is shown to be simple and reliable for acoustic identification.

2.3.5 *Dynamic Fourier Analysis*

Dynamic Fourier analysis is useful for the non-stationary time series analysis, by giving a local-time representation of the spectrum (Shumway and Stoffer 2006). In some cases, the focus of the study is about the non-stationarity of a time series, i.e. the researcher may interest in the local behavior of the process, instead of the global behavior of the process. The dynamic model starts with the creation of subsamples (blocks) of the time series data. Then, the spectral density on each block is computed separately. Finally, the spectral densities are plotted together in a 3D plot. Mathematically, the local Fourier transforms of the time series x_t can be expressed as

$$d_{j,k} = n^{-1/2} \sum_{t=1}^{n} x_t \psi_{j,k}(t)$$

where

$$\psi_{j,k}(t) = \begin{cases} \left(n/m\right)^{1/2} h_t e^{-2\pi i t j/m} & t \in [t_k + 1, t_k + m] \\ 0 & otherwise \end{cases}$$

where h_t is a taper, m is some fraction of n, j indexes frequency, and k indexes the time shift of the transform. The transforms here are based on local sinusoids, i.e. the tapered cosines and sines that have been zeroed out over various regions in time.

It is increasingly popular in medicine to use the frequency-domain analysis of short-term heart rate variability (HRV) measurements (Hartikainen et al. 1998). Spangl and Dutter (2007) used overlapping windows to access the heart rate variability in the frequency domain. Robust dynamic Fourier analysis of the original short-term HRV data was conducted. As another application example, the dynamic Fourier analysis can also be used for the time series analysis of the explosion and earthquake series (Shumway and Stoffer 2006). A Fourier analysis is performed on a short section of the time series data. Then, the section is shifted, and a Fourier analysis is performed on the new section, and so on until the end of the time series. In this way, the spectral behavior of the signal evolving over time can be obtained.

2.3.6 Random Coefficient Regression

Random coefficient regression (RCR) is known to be robust against data that are not missing completely at random (Laird and Ware 1982). The model can handle multivariate ridge-type regressions and inversion problems, and the usual random effects analysis of variance in the frequency domain is a special case in the RCR. Different from the regression models discussed in the above sections, the regression vector of RCR is regarded as unknown stochastic signal. There are many cases in time series analysis that this assumption is more natural. Mathematically, a RCR model can be expressed as (Feuerverger and Vardi 2000):

$$Y_i = A_i + B_i X_i,\ i = 1, \ldots, n$$

where the given time series (X_i, Y_i) are generated from this model, which is with no explicit error term. The unobserved coefficients (A_i, B_i) are assumed to be independent random vectors generated from some unknown distribution with bivariate density f_{AB} (a, b). The task is to estimate f_{AB} from the observed (X_i, Y_i)'s, and the problem of estimating the regression function is often called deconvolution in the engineering literature. Feuerverger and Vardi showed that the nonparametric likelihood function for the RCR problem is the same as that of a positron emission tomography (PET) image reconstruction problem with a suitably specified conditional probability of detection function. With this relation, many of the statistical methods for image reconstruction in PET, like convolution backprojection, Bayes methods, penalized likelihood, and smoothed EM algorithms can be modified for the RCR problem.

RCR is not only of interest in econometrics, but has been applied in other fields as well. Corbett and et al. (2003) built a RCR model to estimate growth curve parameters for the serum glucose trait. The resulting parameters were then used as phenotypes for a variance-components based linkage analysis method. It is shown that the simulated time series is nearly ideal for examination using the RCR model, and the RCR model can virtually recover the model used to simulate the glucose level time series. The RCR growth parameter is proved much more powerful at discovering the location of loci affecting the growth curve for serum glucose levels than the two time-point slope phenotype.

2.3.7 Discrete Fourier Transform

Discrete Fourier transform (DFT) is very useful because they can reveal periodicities in input data as well as the relative strengths of any periodic components (Weisstein 2009). DFT is a specific kind of Fourier transform, with discrete inputs which are often created by sampling a continuous function, like the human voice. The model decomposes a sequence of values into components of different frequencies. Mathematically, the DFT is defined by

$$X_k = \sum_{n=0}^{N-1} x_n e^{-i2\pi k \frac{n}{N}} \quad k = 0,\ldots,N-1$$

where the time series $x_0,\ldots, x_{N-1}$ are complex numbers. As there are N outputs X_k and each output requires a total of N terms, it takes $O(N_2)$ operations as a whole, where O denotes the upper bound of the computational complexity. The inverse discrete Fourier transform (IDFT) is defined by

$$x_n = \frac{1}{N}\sum_{k=0}^{N-1} X_k e^{i2\pi k \frac{n}{N}} \quad n = 0,\ldots,N-1$$

As the inputs for the DFT is a finite sequence of numbers, the model is very suitable for processing information stored in computers, and is widely used in signal processing to study the frequencies in the sample signal. In practice, the DFT is commonly computed much more efficiently using a fast Fourier transform (FFT) algorithm. The FFT algorithms only require $O(N\log N)$ operations. Currently the most common FFT is the Cooley–Tukey algorithm, which is a divide and conquer algorithm that recursively breaks down a DFT of any composite size $N = N_1N_2$ into many smaller DFTs of sizes N_1 and N_2 (Cooley and Tukey 1965).

Chapter 3
Advances in Innovative Computing Paradigms

Abstract In this chapter, it is the brief introduction to the innovative computing paradigms, the advances in the technology and the outline of the recent works of the innovative computing projects. There are different techniques to extract information from various kinds of datasets. In Section 3.1, it is about the research advances in computing algorithms and databases, covering topics like knowledge extraction, data mining algorithms, quantum computing, and DNA computing. In Section 3.2, the focus is on the advances in integration of hardware, systems and networks. Topics like innovative hardware system, graphics processing units, visual exploration, network interoperability, and code optimization are discussed. Section 3.3 is about the advances in Internet and grid computing. Updates about distributed computation, large-scale collaborations over the Internet, pooling of computer resources, and knowledge metadata systems are presented. The advances in visualization, design and communication are described in Section 3.4. Section 3.5 is about the advances of innovative computing for time series problems, like retrieval, automatic classification, clustering, and automatic monitoring of time series. In the last section, it is illustrated how to build an innovative computing algorithm for some simulated time series. Then, in the next three chapters, innovative computer algorithms are built for some real time series data in business, biology and physics.

Keywords Innovative computing • Computing algorithms • Integration • Data sharing • Visualization • Time series

3.1 Research Advances in Computing Algorithms and Databases

3.1.1 Knowledge Extraction Methods

There are different techniques to extract information from various kinds of datasets like structured data, electronic text, and data of natural language etc. For example, as the amount of electronic texts has been increasing dramatically, various methods

S.-I. Ao, *Applied Time Series Analysis and Innovative Computing*,
Lecture Notes in Electrical Engineering 59, DOI 10.1007/978-90-481-8768-3_3,

have been proposed to reduce human workloads to extract the knowledge. It is an active research area to extract knowledge automatically from large-scale text datasets. Information extraction is a popular field in this area. The systems focused on searching instances of pre-specified relations and can recognize entities and relations within text corpora (Banko and Etzioni 2007). Unsupervised methods have also been developed to exploit the large volumes of web pages and can extract structured knowledge from unstructured text. Etzioni et al. (2005) designed the KNOWITALL system that can scale itself to extract knowledge from millions of web pages with diverse size and relationships. The KNOWITALL is based on a boot-strapping approach and utilizes a set of domain-independent extraction patterns to label its own training examples. Banko and Etzioni (2007) proposed a lifelong learning method which used earlier learned knowledge to guide subsequent learning. Their lifelong learning agent can discover now concepts and compose abstract domain knowledge with a precision of 78%.

3.1.2 Exploiting Large Complex Databases

Knowledge discovery in databases (KDD) refers to the extraction of interesting and useful knowledge from real-world databases (Fayyad et al. 1996). Traditionally, KDD methods have been applied to relatively small datasets of typically less than several thousand tuples. With the advances in computer technology, nowadays huge amounts of data can be stored in data warehouses. Databases that integrate large amount of data from heterogeneous information resources lead to large complex databases. In order to exploit these information resources, new KDD systems are developed to cope with the increasing amount of data. A serious difficulty for the application of KDD algorithms to large complex databases is the scalability problem, which may lead to excessive processing time (Cohen 1995). Parallelism is used to speed up the queries that otherwise would takes a long time to process. It can keep the processing times constant, even if the amount of data items increases (Reuter 1999). Query optimizers were designed when optimizing complex queries. Tao et al. (2003) proposed a new similarity-based optimization technique. It can identify groups of similar subqueries in a complex query, and the optimization result within each group in the query was shared.

3.1.3 Neural Computing Algorithms

Among the different tasks that machine learning tools can handle, one popular task is to filter the noises of the source data and then to made prediction basing on the extracted patterns of the source data. The neural network has been found to perform this filtering and prediction capability well. The network can first extract the vital signals and information from the source data. Then it can predict what the future signals and information will be, based on some function approximation. The filtering and prediction capability of the neural network have enabled it to become a popular advance tool for the time series prediction, e.g., the financial prediction of the future index, stock share prices, currency rates, the weather forecast, the traffic control forecast and the medical analysis.

The study of the neural network began after the Warren McCulloch and Walter Pitts have proposed the first mathematical model of a single idealized biological neuron in 1943 (McCulloch and Pitts 1943). The model has been known as McCulloch–Pitts model, which consists of a single neuron that receives the input signals and sums them up with different weights. Then, newer models like the Perceptron by Frank Rosenblatt (1958) and the ADALINE by Widrow (1959) were developed.

Since these earliest works on the neural network, there have come many other neural network models that made use of the neuron concept. The network models that utilize more than one neuron and contain no feed-back paths within the network are given the term feedforward networks. In the feedforward network, there are the single-layer feedforward networks, which consist of the input layer and output layer only, and the multi-layer feedforward networks, which consist of the input layer, hidden layer and the output layer. In our research, we have utilized the multi-layer feedforward network for building our hybrid models for continuous microarray time series analysis.

3.1.4 Fuzzy Computing Algorithms

Fuzzy systems are based on fuzzy logic, which was first proposed by Lotfi Zadeh in 1965. Fuzzy logic is a form of multi-valued logic that deals with approximate reasoning (Zadeh et al. 1996). The degree of truth of a statement is not restricted to the two truth values {true, false}. Instead, it can range between 0 and 1. For example, in a 1,000-ml bottle there are 200 ml of coke, one might define the bottle as being 0.2 full and 0.8 empty, for the two fuzzy sets, Full and Empty. The fuzzy set theory defines the fuzzy operators on these fuzzy sets. A difficulty with the fuzzy systems is that the appropriate Fuzzy Operators may not be known in advance. Ghazavi and Liao (2008) proposed three fuzzy modeling methods including the fuzzy k-nearest neighbor algorithm, a fuzzy clustering-based modeling, and the adaptive network-based fuzzy inference system for medical data mining. The proposed methods were applied to the medical data of the Wisconsin breast cancer dataset and the Pima Indians diabetes dataset. Akdemir (2008) proposed a new method based on combining principal component analysis (PCA) and adaptive network-based fuzzy inference system (ANFIS) to diagnose the optic nerve disease from visual-evoked potential (VEP) signals.

3.1.5 Evolutionary Computing Algorithms

Evolutionary computation techniques are often inspired by the biological systems of evolution (Jong 2006). Evolutionary computing involves the iterative steps of the growth or development in a population. In the iterative process, the population is selected in a random search to achieve the target goal. It includes the metaheuristic optimization algorithms like genetic algorithms, evolutionary programming, evolution strategy, genetic programming, ant colony optimization and particle swarm optimization. Lamers et al. (2008) proposed a hybrid system of artificial neural networks trained via evolutionary computation for predicting viral co-receptor usage. The results show identification of R5X4 viruses with predictive accuracy of 75.5%. Ritchie et al. (2007) applied the multifactor dimensionality reduction (MDR) and grammatical evolution neural networks (GENN) to three data sets from GAW15 for the analysis of rheumatoid arthritis. Rowland (2003) proposed an approach to model selection in supervised learning with evolutionary computation with applications to metabolite determination and to disease prediction from gene expression data. To improve the diagnosis, prevention, and treatment of common, complex human diseases, Moore et al. developed a hierarchical dynamic systems approach based on Petri nets for generating biochemical network models that are consistent with genetic models of disease susceptibility. An evolutionary computation approach – grammatical evolution, is used as a search strategy for optimal Petri net models (Moore et al. 2005).

3.1.6 Quantum Computing Algorithms

Quantum computing algorithms are a fascinating and rapidly developing area, involving topics from physics, mathematics and computer science (Pittenger 2000). Quantum computers (Vandersypen et al. 1999) are different from traditional computers based on transistors. The memory of a classical computer is consisted of bits, where each bit represents either a one or a zero. A quantum computer has a sequence of qubits, where each qubit represents a one, a zero, or any quantum superposition of these. These qubits are manipulated with a fixed sequence of quantum logic gates, called a quantum algorithm. Currently, quantum computing is still in its infancy, but the quantum computers have the potential to solve certain problems like the integer factorization and Shor's algorithm much faster than the traditional computers (Lenstra 2000).

3.1.7 Swarm-Based Computing Algorithms

Swarm-based computing was introduced by Beni and Wang (1989), based on the collective behavior of decentralized, self-organized systems. Swarm-based systems are typically consisted of a population of simple agents interacting locally with each other and with the environment. These interactions can lead to the emergence of intelligent global behavior. Natural examples include for example the ant colonies, bird flocking and bacterial growth etc. Example algorithm of the swarm-based computing included ant colony optimization, particle swarm optimization and stochastic diffusion search. In the ant colony optimization, the simulation agents – artificial ants – find the optimal solutions by searching through a parameter space representing all possible solutions (Dorigo and Stützle 2004). Particle swarm optimization is a global optimization algorithm, where particles move through the solution space and accelerated towards those particles within their communication grouping of better fitness values (Parsopoulos and Vrahatis 2002). Stochastic diffusion search is an agent-based probabilistic optimization method, with a positive feedback mechanism to ensure that, over time, a population of agents stabilizes around the global-best solution. Swarm-based computing has been applied in number of applications, like the vehicle control, planetary mapping, crowd simulation, and ant-based routing, etc.

3.1.8 DNA Computing Algorithms

DNA computing was initiated by Adleman (1994), who showed a proof-of-concept use of DNA as a form of computation to solve the seen-point Hamiltonian path problem. It is a fast developing interdisciplinary computing area that uses DNA and molecular biology, instead of the silicon-based computer technologies. As DNA uses adenosine- triphosphate as fuel to allow ligation, its power consumption is much lower than traditional silicon computers. In DNA computing, the many different molecules of DNA can try many different possibilities at once. For a number of problems, the DNA computer runs faster than the other traditional computers.

3.1.9 Theoretical Modeling and Simulations

The modeling and simulation approach is to simulate an abstract model of a particle model. It can help to understand the interaction among the parts of a system, and is useful for many natural systems astrophysics, biology, human systems in economics, and social science, to gain insight into the operation of those systems. Simulations are often used as a substitution for, or adjunct to, modeling systems where simple closed form analytic solutions are not possible or very difficult to obtain. A simulation can be considered as a computerized version of the model, and it is run over time to study the implications of the defined interactions. In the modeling and simulation approach, a model is developed and then simulated. Knowledge was learned from the simulation and then the model was revised and simulated again. The iterations continue until satisfactory level of knowledge is obtained.

3.2 Research Advances in Integration of Hardware, Systems and Networks

3.2.1 Innovative Experimental Hardware System

Innovative experimental hardware system includes quantum computer, chemical computer, DNA computing, optical computer, and spintronics based computing etc., with quantum and DNA computing described in the above section. In a chemical computer, the computations are performed by naturally occurring chemical reactions.

Adamatzky (1997) showed simple logic gates using reaction-diffusion processes, and how a cellular automation can perform computation. A potential advantage of the technology is that a cup of chemical is a relatively simple, cheap and stable component to produce when compared with the traditional microprocessor. Another potential is that the chemical computer may be able to handle computation much faster than a traditional computer. A traditional computer uses electric current to do the computation, whereas an optical computer is a devise that utilize the photons in visible light or infrared beams to perform digital computation. Unlike electric current, light does not create significant amounts of heat during the computation. Spintronics technology utilizes the intrinsic spin of the electron, its associated magnetic moment, and its fundamental electronic charge, in solid-state devices. The epintronics-enabled devices like GMR and TMR sensors can increase the sensitivity of the read head of the hard drives for the information storage, and thus contribute to the exponential growth of the storage density of the hard drives. Its other advantages include the potentially lower power consumption and a small footprint than traditional electrical devices.

3.2.2 Data-Acquisition Devices

Data-acquisition devices convert physical parameters to electrical signals that can be manipulated by a computer. It usually involves the conversion of analog signals and waveforms into digital values and the processing of the values to obtain desired information. The data-acquisition devices act as the interfaces between the signal and a computer. Intelligent data-acquisition devices, like intelligent stand-alone loggers and controllers, can be operated from a computer, yet the devices can also operate completely independent of the computer.

3.2.3 Interaction Devices for Visual Exploration

The interaction devices provide a way for computer users to accomplish their tasks, and act as interfaces for the communication between the systems and the users. The interfaces include media for visual representation of the system state, and media for visual user input or manipulation. The visualization is an interdisciplinary study based on research in human–computer interaction, computer science, graphics, visual design and psychology, and becomes an important component in scientific research (Bederson and Shneiderman 2003). It is about the visual representation of large-scale collections of information. Visualization techniques are interaction techniques that involve data exploration with emphasis on output presentation. Specific techniques include for example like cladogram, color alphabet, dendrogram, halo, heatmap, and multidimensional scaling etc. Visual displays can help the users to explore the information in visual form like 3D outputs. The visual representations can talk advantage of the human eye's broad bandwidth pathway into the mind to allow users to explore visually large amounts of information at one time. Head-mounted displays and cave automatic virtual environment are examples of a fully-immersive visual display. Monitors and workbenches are example of semi-immersive displays. Fully-immersive visual display is where user can see only the virtual world, while semi-immersive display allow users to see both the virtual world and the real world.

3.2.4 *Graphics Processing Units and Co-Processors for Innovative Computing*

Co-processors are computer processor supplementing the functions of the primary processor. The graphics processing unit (GPU) in the form of graphics card is a common modern PC co-processor. GPUs are specialized processor that offloads 3D graphics rendering from the microprocessor. Not only are the GPU very efficient at manipulating computer graphics, they are an attractive platform for general purpose computation, as the highly parallel structure of the GPUs make them more effective than general-purpose CPUs for many complex algorithms. Current developments in GPUs include support for programmable shaders to manipulate vertices and textures. Interpolation techniques to reduce aliasing, and very high-precision color spaces. Scientists have also increasingly conducted research on the applications of GPUs for non-graphical calculations. For example, there have been interests in applying the GPU for evolutionary computation. The performance can be improved by running the single active program simultaneously on many example problems in parallel. Substantial acceleration can also be achieved by transferring the problems to the GPU and interpreting them there (Garcia et al. 2008).

3.2.5 *Networking and Interoperability*

In networking, interoperability refers to the continuous ability to transfer data between interconnected networks providing the level of quality expected by the end user customers. For social networking in Web 2.0, the interoperable framework can enable the applications to share profiles and data across networks to open further Web 2.0 possibilities. To achieve the interoperability between networks of specialized protocols, Fall (2003) proposed a network architecture and application interface structured around optionally-reliable asynchronous message forwarding, with limited expectations of end-to-end connectivity and node resources. Traditionally packet networks achieve interoperability by standardizing the syntax and semantics of packets. For example, internet routers all support the agreed IP specifications. Tennenhouse and Wetherall (2007) proposed instead the standardization of the computation model, i.e., the instruction set and resources available to capsule programs. Then, network interoperability can be achieved at a higher level of abstraction, instead of standardizing the computation performed on very packet.

3.2.6 Code Optimization and Integration

As the high-level languages (HLLs) like C (and C++) can greatly lowers the cost of development and maintenance of embedded systems, HLLs are replacing assembly language in these systems. However, current compilers for fixed-pointed digital signal processing (DSP) microprocessors generate poor code. Thus programming in HLL can incur significant code performance and code size penalties. Liao et al. (1998) addressed this problem by proposing optimal and heuristic algorithms that could determine an instruction schedule simultaneously optimizing accumulator spilling and mode selection. There are also researches to explore automated, distributed computing frameworks for integrating disciplinary analysis tools. These design frameworks allow collaborative design teams to take advantage of distributed expertise and existing legacy codes, while preserving some of the automation and optimization capability of monolithic synthesis tools and simple subroutines (Olds and Steadman 1998).

3.3 Research Advances in Internet, Web and Grid Computing

3.3.1 Distributed Computation and Data Sharing

In distributed computation, the system consists of autonomous computers that communicate through a computer network. There are several basic architectures of distribute computing: client-server, distributed objects, loose coupling, tight coupling, N-tier architecture, and 3-tier architecture. Another aspect of the architecture is about the method of communication and coordinating work among concurrent processes. It is a challenging problem to study communication and data sharing in highly dynamic distributed environment. Lynch and Shvartsman (2003) proposed the new dynamic atomic shared-memory service for message-passing systems, where the set of owners of data can be changed dynamically and concurrently with the ongoing read and write operations. Common middleware frameworks such as CORBA, DCE and Java/JINI support the construction of distributed systems from components. Their specification capability is restricted to the formal definition of interfaces and informal descriptions of behavior. In Lynch's works, the proposed middleware RAMBO differs from this common practice and is specified in a compositional way, such that the system properties can be inferred from corresponding properties of the system components.

3.3.2 Large-Scale Collaborations over the Internet

Traditional communication methods like hardware videoconference can only serve the communication between a few persons, and may not satisfy the requirements of large-scale collaborations like scientific collaborations. Collaboration often presents unforeseen social and organizational challenges (Olson et al. 2008). Joslin et al. (2000)

proposed a networked virtual environment system of the ability to transfer data across the Internet so that geographically distant users can collaborate with each other. The system can provide realistic virtual actors and sets of applicable high-level actions in real-time. Jin et al. (2004) introduced a platform for large-scale multimedia and data collaboration over the Internet. Their prototype platform contained multimedia communication and basic data collaboration function.

3.3.3 *Grid Computing*

Grid computing refers to the combination of computer resources from multiple administrative domains applied to a common computational task. Specific software is used to divide and apportion pieces of the task among computers. Grid computing differs from conventional cluster computing by its more loosely coupled, heterogeneous, and geographically dispersed properties. A difficulty in grip computing is that the designer must have measures to prevent malfunction in any participating computer from producing erroneous results. Various middleware projects like BOINC (boinc.berkeley.edu) have been creating generic infrastructure to allow projects to harness a particular associated grid, or to set up new grids.

3.3.4 *Pooling of Remote Computer Resources*

Pooling of the remote computer resources can group together the computers for the purpose of maximizing the productivity of the resources to the users. The pooling can make a collection of computer resources behave like a single pooled resources. The main advantages of resource pooling include increased robustness against component failure, better ability to handle localized surges in traffic, and maximized utilization (Wischik, et al. 2008). Rajan et al. (2008) deployed the computing resources by pooling and built a virtual grid over a local area network. Software module was proposed to achieve the integrating of the pooled resources. Pooling techniques have already been used commercially in the content distribution networks of Akamai and Google etc. In the network, a large amount of redundant servers are distributed in multiple data centers. The resources like CPU cycles, bandwidth and reliability are pooled with the DNS load-balancing and hardware load balancing at each data center.

3.3.5 *Integration of Knowledge Metadata Systems*

Metadata can describe the data conceptually and qualitatively in formats like text, voice, or image, to facilitate the understanding and management of data. Despite the recent advances in management support systems (MSS), data warehouses are often built to implement an integrated and consistent collection of decision relevant numerical data only (Kimball 1996). Supplementary qualitative information for effective analysis and evaluation is often missing (Fedorowics 1996). Many metadata subsystems may focus only on the management of the technical processes of loading and cube generation. Devlin (1997) demanded the integration of different types of metadata across boundaries of specific methodologically divided management support systems. The metadata component can be outsourced by the MSS-components like data warehouses, business intelligence tools, and repositioned on a superior level. Nemati et al. (2002) supposed important adaptations and new considerations about current architectures to model metadata to match knowledge representation on the domain of knowledge warehouses.

3.4 Research Advances in Visualization, Design and Communication

3.4.1 Novel Solutions to Visualization and Communication Challenges

Knowledge visualization and interactivity is an essential issue with regard to organizational communication. In a recent DaimlerChrysler research project, the interactive visualization was used to support the communication between several different management levels and professional backgrounds (Klein 2005). For scientific projects, it is also a challenging task to have effective communication within the scientific team, as terabytes of data may be produced in a single scientific project. The communication challenges for the scientific projects include the tasks of keeping scientists updated and informed, and organizing the scientific results. Visualization techniques and tools like ParaView, VisIT, POVRay, volume renderin, particle rendering, are used in real-time analysis and filtering of the data. To address the need of real-time analysis, ParaView (www.paraview.org) has parallel capabilities and can supports additional computation. It has simplified interface and TeraGrid visualization gateway. Particle rendering is about the filtering of data like geometric extraction and data cuts.

3.4.2 Displaying of Complex Information

It is important to display the complex information clearly, precisely and efficiently. Good graphical displays should include characteristics like inducing the viewer to think about the substance, avoiding distorting the data, making large data sets coherent, revealing the data at several levels of details, and closely integrating with the statistical and verbal descriptions of the data set (Tufte 1992). In commercial applications like flight traffic control, good three-dimensional flight display can increase the efficiency of remote sensing flight operations by providing enhanced guidance and situational awareness on straight and curved flight paths. Barrows and Powell (1999) proposed a prototype system with differential GPS, inexpensive graphics hardware, and a flat-panel display, with significant advantages over conventional instruments. The system is able to guide the pilot on complex paths. Displaying of complex information is also useful in scientific works. Tudela et al. (2003) proposed a method to display complex Fresnel holograms by adding the information displayed on two analogue ferroelectric liquid crystal spatial light modulators. The real and imaginary parts were displayed on each panel, and the object was reconstructed at that distance from the modulators by simple propagation of light.

3.4.3 Escaping Flatland

In the classic book Flatland, Abbott (1899) described a world of two dimensions, where everything looks like a line in the two-dimensional world of Flatland. In the work of Tufte (1990) about "Escaping Flatland", it was called for the increasing of data density in visual displays. Tytell et al. (2008) showed for instances that one needed to escape Flatland to examine all three dimensions of fishing swimming, as fish fins could move in a complex three-dimensional manner. An understanding of the three-dimensional body shape and fin motions is important for explaining the mechanics of swimming. The traditional two-dimensional representations of fish locomotion can often be misleading. Visualization tools like Knowledge Space Visualizer (KSV) have been developed for information visualization. KSV employed a force directed layout algorithm to minimize the distortion of the relationships as viewed in two dimensions. Results of the visualizations could be fed back into the data source environment to promote further building on of analyses by other researchers. For some scientific researches, it is important not to be restricted within the studies of two-dimensional components.

3.4.4 Systems Integration for High Performance Image Processing

Signal process applications in areas such as image, audio, and speech processing often require high-sample rates and distributed computation. Designers can achieve accelerated system clock rate with heavily pipelined implementation of functional blocks and inter-block communication (Tessier and Burleson 2001). These implementations have showed substantially better performance than the standard methods. For many scientific projects, it is an important step to analyze large and distributed image datasets in order to help understand the characteristics of the systems concerned. Hastings et al. (2003) designed a toolkit to allow the rapid and efficient development of biomedical image analysis allocations in a distributed environment. The toolkit was shown to be able to achieve good performance by enabling combined task and data parallelism in a chain of processing operations. For real-time image processing, specialized data paths and pipelining are typically required. A number of research projects have been conducted to address the issues involved. For instances in Athanas and Abbott (1995), a set of image processing tasks was mapped to the Splash II platform. Methods like Gaussian pyramid-based image compression, image filtering with 1D and 2D transforms, and image conversion were studied.

3.4.5 Representation of Uncertainties

Widely used approaches for representation of uncertainties include interval mathematics, classical set theory, fuzzy theory, and probabilistic analysis (Klir 1994). The interval mathematics can address data uncertainty due to the imprecise measurement and the existence of alternative methods to estimate model parameters. It objective to estimate the bounds on various model outputs based on the bounds of the model inputs and parameters. Symbolic computational packages such as Mathematica and Maple support interval arithmetic. In classical set theory, uncertainties, like diagnostic, predictive and retrodictive uncertainties, are expressed by sets of mutually exclusive alternatives in situations where one alternative is sought. Fuzzy theory can be applied in uncertainty analysis for systems where the uncertainty is due to vagueness rather than due to randomness alone (Evans et al. 1986). Fuzzification is used to generalize any specific theory from a crisp to a continuous fuzzy form. Probabilistic analysis is a widely used method for describing the uncertainties in physical systems. In the probabilistic approach, the probabilities associated with events are used to describe the uncertainties. Different modeling techniques have been used to solve the uncertainty in real-world problems. For instances, Cristina et al. (2009) proposed a spatial modeling of a decision-making process in water management, taking into accounts the associated uncertainties. The fuzzy logic is used to handle the subjective data as well as the associate uncertainties. Tao (2004) designed the robust controllers for uncertain systems with the linguistic uncertainties represented by fuzzy sets. A robust sliding mode controller was applied to stabilize the uncertain system if the sliding coefficient conditions are satisfied.

3.4.6 Informative Graphics for Scientific Communication

Scientific communication is an interdisciplinary field, where both written and visual languages are used. When the relationships among variables are complex or time-series data, graphical display is often a helpful tool to communicate the results.

It can help to clarify the ideas. For example, the graphic representation of the orbit of a spacecraft, or the visual representation of the structure of deoxyribonucleic acid (DNA) as a double helix, is visually stunning (Peck 1995). With the technological advances, huge amount of data is produced, and significant changes have been taking place in visual communication. Scientists are using more interpretive graphic art than ever to convey their ideas and results. New software like Adobe Illustrator and Aldus Freehand are manufactured to enhance graphic production. New visual techniques are developed to better understand and communicate the experimental and computational results. For instances, in fields like microbiology, data are often displayed in tables or graphs. And numbers are often transformed into colours in order to discover the underlying complex relationship between the data (Harmon and Gross 2007). Appropriate usage of graphics can help to obtain a good approximation to the available information.

3.5 Advances and Applications for Time Series Problems

3.5.1 Efficient Retrieval of Similar Time Series

With the advances in the computer and data storage technology, massive data are generated. Many times the data are time-varying, for instances data from biomedical or astrophysical sensors. One ongoing research direction in time series data mining is to develop algorithms that can identify similar temporal patterns in a collection of time series. The identification and retrieval of similar time series pattern may lead to the discovery of hidden information in time series database (Keogh et al. 2001). Saeed and Mark (2006) proposed a novel temporal similarity metric based on a transformation of time series data into an intuitive symbolic representation. After the symbolic transformation, traditional information retrieval algorithms based on a vector–space model can be utilized. The proposed algorithm is shown to be computational efficient and can assess the similarity between multi-dimensional time series. Rafiei and Mendelzon (1998) proposed an improvement of the known indexing technique based on Discrete Fourier Transform (DFT) for fast retrieval of similar time sequences. The last few Fourier coefficients in the distance computation were used without storing them in the index. This technique could accelerate the search time of the index by more than a factor of two.

3.5.2 Automatic Classification of Time Series Sequences

Segmentation or classification of time series into discrete classes is an important task in many fields (Lenser and Veloso 2005). Lenser and Veloso proposed an improved state-based prediction algorithm for time series. The algorithm can predict a multi-modal distribution over future values and forms the basis for labeling part of a time series with the underlying state. It is robust to fluctuations in signal strength and noises, and can identify sudden changes in the robot's environment allowing the robot to respond intelligently. Molina et al. (2009) proposed a process classification model based on time series data segmentation. The segment classification method can identify micro behaviors and the way the process is transforming. The approach has been successfully applied in various engineering domains like failure predictions in oil process plant, trajectory reconstructing in air traffic control and identification of interaction scenarios in mobile robotic environment. Dietrich et al. (2004) studied the automated classification of bioacoustic time series based on the combination of global and local decisions. The subject in the study was the cricket songs from Thailand and Ecuador. A fuzzy-*k*-nearest-nearest-neighbour classifier was used to classify the time series, basing on their local and global features.

3.5.3 *Time Warping Algorithms*

Dynamic time warping (DTW) is used to measure the similarity between two time sequences. It can allow a non-linear mapping of one sequence to another by minimizing the distance between the two (Ratanamahatana and Keogh 2005). DTW method can be applied for various tasks in time series problems like clustering, classification and anomaly detection etc. An advantage of DTW is its flexibility to allow two sequences that are similar but locally out of phase to align in a non-linear way. The time complexity of DTW is $O(n^2)$. For the problems like this study of a few hundred sequences, each with a few hundred data points, the CPU cost of DTW is simply not an issue. Otherwise, speed improvements like smaller warping windows can be used to speed up the DTW calculations.

The pattern similarity between the autocorrelation sequences of the light curves is determined by aligning two autocorrelation sequences A and B, with distortion $D(A, B)$, which is two of the autocorrelation sequences of the light curves. The objective is to have a decision rule that can choose the alignment with the smallest distortion:

$$R^* = \arg\min D(A, B)$$

Dynamic time warping is employed to obtain the best possible alignment warp between A and B, along with the associated distortion R^*. Mathematically, let the two sequences be represented as $A = \{a_1,\ldots,a_N\}$ and $B = \{b_1,\ldots,b_M\}$. An alignment warp, $\varphi = (\varphi_a, \varphi_b)$, aligns A and B through a point-to-point mapping of length K_φ

$$a_{\varphi_a(k)} \Leftrightarrow b_{\varphi_b(k)} 1 \le k \le K_\varphi$$

The overall distortion $D_\varphi(A,B)$ of a particular alignment warp ϕ is the sum of the local distances between elements $d\left(a_{\varphi_a(i)}, b_{\varphi_b(j)}\right)$:

$$D_\varphi\left(A,B\right)=\sum_{k=1}^{K_\varphi} d(a_{\varphi_a(k)}, b_{\varphi_b(k)})$$

With the boundary constraints $\varphi_a(1) = \varphi_b(1) = 1$, $\varphi_a(K) = N$, $\varphi_b(K) = M$, and the monotonicity property $\varphi_a(k+1) >= \varphi_a(k)$, $\varphi_b(k+1) >= \varphi_b(k)$. The optimal alignment is the one that minimizes the overall distortion:

$$D(A,B) = \min_\varphi D_\varphi\left(A,B\right)$$

As an example, for two sequences of signal 1: [3 4 10 6 8] and signal 2: [9 10 12 13 10 17], the computed optical alignments are [3 4 10 10 10 10 6 8] and [9 9 10 12 13 10 10 17] respectively, as shown in Fig. 3.1.

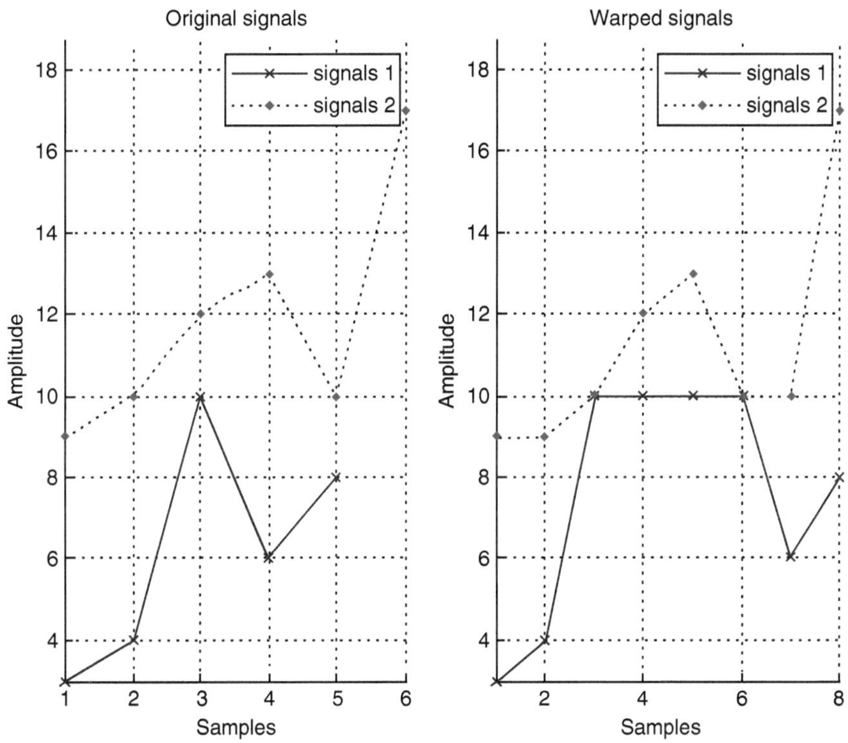

Fig. 3.1 The original signals (*left*) vs. the warped signals by the dynamic time warping method (*right*)

3.5.4 Time Frequency Clustering of Time Series Datasets

Clustering of multivariate time series data can find the groups of data that share similar characteristics. The knowledge obtained from the clustering can be helpful for further analysis like process improvement and fault diagnosis. The high dimension space of long time-series data and the missing data often pose difficulties for the clustering of time series. To solve these difficulties, Wang et al. (2006b) proposed a method to cluster the time series based on their global features extracted from the time series. The feature measures are obtained by applying statistical operations that best capture the underlying characteristics like trend, seasonality, serial correlation, skewness, kurtosis, and nonlinearity, and can be fed into arbitrary clustering algorithms like neural network or hierarchal clustering algorithms. Singhal and Seborg (2005) proposed a new clustering methodology based on the calculation of the degree of similarity with principal component analysis and distance similarity factors. The similarity factors are based on the Mahalanobis distance between the datasets. The simulation results of two nonlinear dynamic systems support the effectiveness of the method. Subsequence time series clustering is a popular discovery technique from time series data, but its mathematical methodology has not yet been fully studied. Fujimaki et al. (2008) conducted, for the first time, a theoretical analysis of subsequence time series clustering from a frequency-analysis iewpoint. A clustering algorithm using a phase alignment preprocessing was also proposed to obtain meaningful clustering results.

3.5.5 Enhanced Representation for Complex Time Series

Good representation for complex time series can allow efficient computation on data, and extract higher order features. Methods like Fourier transformations (Faloutsos et al. 1994), relational trees (Shaw and DeFigueiredo 1990) and envelope matching (Agrawal et al. 1995) have been proposed for the representation. Nevertheless, all the above methods have shortcoming like sensitivity to noise, lack of intuitiveness, and the need to fine tune many parameters (Keogh and Pazzani 1998). Keogh and Pazzani proposed an extended representation of time series that allowed fast, accurate classification and clustering and with the ability to explore time series in a relevance feedback framework. The representation consists of piece-wise linear segments to represent shape and a weight vector that records the relative importance of each individual linear segment. This representation can enable the user to define a variety of similarity measures tailored to specific domains. Another representation for the time series is the symbolic representation, where real valued time series is transformed into symbolic representations. The representation may enable the researchers to avail of the data structures. Lin et al. (2003) proposed a new symbolic representation of time series that have innovative features like dimensionality/numerosity reduction, and distance measures that are defined on the symbolic approach. The representation is shown to be able to allow the real value data to be converted in a streaming fashion, with only an infinitesimal time and space overhead.

3.5.6 Automatic Monitoring of Large and Complex Time Series

An objective of monitoring the time series is the identification of specific events promptly. Hunter and McIntosh (1620) proposed a promising approach to the detection of events in a complex, multi-channel, high frequency time series data. A software workbench was developed to allow the expert to display the data and to mark up features of interest. Then this knowledge is sued to define the parameters for a pattern matcher. The matcher was run over a set of time series intervals by a new iterative interval merging algorithm. Lin et al. (2004) designed a novel time-series visualization and monitoring tool to aid aerospace analysts. The visualization approach worked by transforming the time series into a symbolic representation, and encoding the data in a modified suffix. The tool can allow a natural transfer of mined knowledge to the monitoring task. Charbonnier et al. (2005) proposed an effective trend extraction procedure for complex time series system monitoring and decision support. The method can extract semi-qualitative temporal episodes online from any univariate time series. The method uses a segmentation algorithm, a classification of the segments and a temporal aggregation of episodes.

3.6 An Illustrative Example of Building an Innovative Computing Algorithm for Simulated Time Series

3.6.1 Description of the Simulated Time Series Problem

In this section, it is illustrated how to build an innovative computing algorithm for some simulated time series. Then, in the next three chapters, innovative computer algorithms are built for some real-world time series data in business, biology and physics. Here, suppose that we are given some time series values simulated from two generators. The two generators use different regression mechanisms to produce the time series artificially. The problem is that we do not know which generator the time series values are from, and the task is to classify the time series values with the correct generator. It can be observed that this problem is a specific type of the general problem of clustering datasets of nonparametric regression models. In the above problem, the datasets are restricted to the type of time series data only. It is illustrated here how to build a hybrid intelligent algorithm of neural network regression and unsupervised fuzzy clustering for clustering these datasets of nonparametric regression models. In the new formulation, (i) the performance function of the neural network regression models is modified such that the fuzzy clustering weightings can be introduced in these network models; (ii) the errors of these network models are feed-backed into the fuzzy clustering process.

This illustrated hybrid innovative computing approach leads to an iterative procedure to formulate neural network regression models with optimal fuzzy membership values for each object such that the overall error of the neural network regression models can be minimized. The testing results show that this hybrid algorithm NN-FC can handle cases that the K-means and Fuzzy C-means perform poorly (Ao 2008c). The overall training errors drop down rapidly and converge with only a few iterations. The clustering accuracy in testing period is consistent with these drops of errors and can reach up to about 100% for some problems that the other classical fuzzy clustering algorithms perform poorly with about 60% accuracy only. The illustrated algorithm can also build regression models, which has the advantage of the NN component, being non-parametric and thus more flexible than the fuzzy c-regression.

3.6.2 *Background of the Methodology*

Hathaway and Bezdek (1993) developed a methodology of switching regression models and fuzzy clustering. Their main idea is to employ a fuzzy clustering for linear or polynomial regression models for data sets. Their approach leads to an iterative procedure to formulate linear or polynomial regression functions with optimal fuzzy membership values for each object such that the overall error of the linear or polynomial regression functions is minimized. Hathaway and Bezdek have pointed out that there are various applications of the switching regression models in economics. And an example from fisheries was illustrated. The sexuality of a fish called halibut is indistinguishable. The mean length of a male halibut depends on its age for a certain range of ages and so does the female halibut. And the problem can be treated as a switching regression problem of two models, one for the male and another for the female. For the example, Hathaway's two models are

$$y = \beta_{11}x + \beta_{12} + \varepsilon_1$$

$$y = \beta_{21}x + \beta_{22} + \varepsilon_2$$

where y = length and x = age. And our proposed approach can give solutions of the form

$$y = f_1(x) + \varepsilon_1$$

$$y = f_2(x) + \varepsilon_2$$

where the functions are to be simulated by two neural network regression models respectively. Menard (2001) extended fuzzy clustering and switching regression models using ambiguity and distance rejects. The main drawback of switching regression approach is that the parameters of the formal generating functions must be known. However, this may not be valid in practice.

The general regression models for representing the dependent variable $Y(t)$ and independent variable $X(t)$ can be written as followed:

$$Y(t) = F_i(X(t)) + \varepsilon$$

for $1 \leq i \leq K$, $1 \leq t \leq T$ where ε are the noises of the data type, K is the number of different regression functions, T is the total number of time points in the variable, and F_i is the ith regression function between $X(t)$ and $Y(t)$. The difficulty of solving this problem is that (i) the regression functions are unknown and (ii) the regression functions are not labeled. For (i), it implies that the form of F_i is unknown and that it may be linear or nonlinear. For (ii), it implies that the clustering of the data $X(t)$ and $Y(t)$ for the regression functions is required. Our aim is to cluster a set of points into two groups, and to construct their corresponding regression functions for this set of points. We described the nonparametric neural network (NN) regression models to remove such parametric models on regression functions. Our intelligent algorithm can provide non-linear non-parameter solutions to the above applications of the switching regression models. Furthermore, it can cluster datasets produced by some underlying generated functions.

Many clustering problems will be solved poorly with common clustering methods, for instance K-means clustering and Fuzzy c-means (FCM). Our tests of applying these clustering methods can give us a clustering accuracy of about 60% only. It is because the K-means and Fuzzy C-means techniques have strict restrictions on the shapes of the clusters being studied (hyper-spherical clusters of equal size for Euclidean distance and hyper-elliptical clusters for Mahalanobis distance) (Jajuga et al. 2002). Even though Gustafson-Kessel and Gath-Geva algorithms have extended the Fuzzy C-means algorithm for shapes of ellipses and ellipsoids (Hoppner et al. 1999), there are still strict restrictions on the clusters' shapes. For the clustering problem in Fig. 3.1 with two centers, the FCM will cluster the left-hand side data points into one group and the right-hand side data into another group. It is clear that this is a very poor cluster decision for this problem. Fuzzy c-regression can solve the problem partially, only when the parameters of the formal generating functions are given. But, in reality, it is difficult to know in advance about the generating functions.

There are studies (Sarkar et al. 1998; Ronen et al. 1998; Boca and Park 1994; Bortolan and Pedrycz 1998) that combine the fuzzy clustering and the neural network for supervised classification purposes. They have applied the fuzzy clustering to the original data and got the membership values for each object in each cluster. This information can serve as the weighting for the neural network output at the performance level in Sarkar's study. Or it is used during the combining of different NN models' outputs in Ronen's work. In Boca's study, they employ the Fourier analysis and fuzzy clustering to extract the signal features for the supervised neural network classification. Bortolan has applied the fuzzy clustering as a preprocessor for the initialization of the receptive fields of the radial basis function neural network for supervised classification.

It should be noted that these studies have employed the hybrid system in the loose format. That is to pass the variables through each component only once. They have reported satisfactory results for problems that are suitable for supervised learning. But, these studies are restricted to problems that suit the Fuzzy C-means clustering and will perform poorly for other problems. Our study differs from these above methodologies in that we deal with unsupervised learning problems instead of their supervised classification problems. Also, we have developed our algorithm in an iterative manner and achieve the hybrid objectives of unsupervised clustering and neural network regressions with our algorithm. We have formulated a hybrid iterative methodology of neural network and unsupervised fuzzy clustering so that the clustering and regression components can supplement each other for further improvement. A general solution for these problems can be found with our new NN-FC algorithm, which can give clustering accuracy of about 100% in our testing period.

3.6.3 Building the Innovative Regression Model

Our illustrated model consists of the neural network and the unsupervised fuzzy clustering components. The neural network is for investigating the regression relationships between the dependent and independent variables, while the unsupervised fuzzy clustering is for clustering the objects into different generating functions. Then, NN regressions are implemented to simulate different generating functions with the modified performance functions that each error is weighted by the output from the FC. The respective performance errors are passed back to the FC. FC component will adjust the membership values for each object in each cluster based on the corresponding errors of that object in each generating NN model.

The idea comes from the fact that, if an object fits a particular NN model well, its error in that model will be much lower than its errors in other NN models. At the same time, when the memberships are more and more close to the corrected generated clusters, it can help the NN to have better simulations for each underlying regression function. These two components work in a hybrid way and form a loop until the overall errors do not show any further significant improvement.

3.6.3.1 Neural Network Regression Models

Neural network is well known for its non-linear capability and is usually employed with the three-layer architecture. The layers are input layer, hidden layer and output layer. The inspiring idea for this structure is to mimic the working of our brain. The mathematical structure for the above neural network structure can be expressed as followed (Principe et al. 2000):

$$y = f\left(\sum_{j=1}^{J} w_j^{(2)} f\left(\sum_{i=1}^{I} w_{ji}^{(1)} x_i\right)\right)$$

where the function f is the activation function of the network, I denotes the number of inputs, J the number of hidden neurons, x_i the ith input, $w^{(1)}$ the weights between the input and hidden layers, $w^{(2)}$ the weights between the hidden and output layers.

Unlike traditional neural networks, we have employed the fuzzy membership results from the clustering as the weighting for each output error of the network. We have adopted the fuzzy clustering instead of the hard clustering. If the hard clustering of membership values 0 and 1 were employed, the network would be trained with these crisp weightings and then optimized with respect to these crisp weights. When we perform the clustering procedure for such data in the next iteration, the objective function value does not improve. The main reason is that the NN regression models fit the data points very well. It is difficult to adjust the weightings to the optimal membership values when we were restricted with the membership values 0 and 1 only.

As said, we have modified the performance function of a typical network:

$$E = \frac{1}{2NM}\sum_{n=1}^{N}\sum_{m=1}^{M}(z_{nm} - t_{nm})^2$$

with our new one:

$$E = \frac{1}{2NM}\sum_{k=1}^{K}\sum_{n=1}^{N}\sum_{m=1}^{M} w_{knm}^{\alpha}(z_{knm} - t_{knm})^2$$

where N is the number of examples in the data set, M the number of outputs of the network, t_{knm} the mth target output for the nth example in the kth cluster, z_{knm} the mth output for the nth example in the kth cluster, K the number of clusters. w_{knm} is the fuzzy membership value for each sample to belong a certain cluster k, and α is the fuzzy weighting exponent. We have developed specific learning laws for this modified NN performance function similar with the neural network part of Sarkar's study. Sarkar et al. have discussed the motivations and advantages of introducing of this fuzzy mean square error term. In brief, with the introduction of the fuzzy mean square error term, the restriction of an input datum belonging to one and only one cluster/class has been removed. It addresses the situations where the datum may belong to more than one cluster. And the training of the networks can be conceptually viewed as a fuzzy constraint satisfaction problem.

The derivation for a simple three-layer network with logistic transfer function is as followed. In a simple three-layer network, the inputs to hidden neuron j can be denoted as $u_j = a_{0j} + \sum_{i=1}^{I} a_{ij}x_i$, and its output as $y_j = g(u_j)$, where the function is the Logisitic transfer function. The inputs to k output are given as $v_k = b_{0k} + \sum_{j=1}^{J} b_{jk}y_j$, and its output is $z_k = g(v_k)$. For the on-line learning case, the mean squared error is given by:

$$\overline{E} \triangleq \frac{1}{2}\sum_{k=1}^{K}(z_k - t_k)^2$$

Here, we have replaced this measurement of error by our new one E^* with the weightings w_k^* for each output k, where w_k^* is obtained from our fuzzy clustering component,

$$E^* \triangleq \frac{1}{2}\sum_{k=1}^{K} w_k^*(z_k - t_k)^2$$

As a result, the updating procedure for the neural network is different from the typical network and will be derived as followed in a similar way as Sarker et al.

Taking derivative of E^* with respect to z_k,

$$\frac{\partial E^*}{\partial z_k} = w_k^*(z_k - t_k)$$

And we can update $\frac{\partial E^*}{\partial b_{jk}}$ as followed:

$$\frac{\partial E^*}{\partial b_{jk}} = \frac{\partial E^*}{\partial z_k}\frac{\partial z_k}{\partial v_k}\frac{\partial v_k}{\partial b_{jk}} = \begin{cases} P_k^*, \mathit{for}\ j = 0 \\ P_k^* y_j, \mathit{for}\ j = 1,\ldots,J \end{cases}$$

where

$$P_k^* = w_k^*(z_k - t_k)z_k(1 - z_k)$$

And $\frac{\partial E^*}{\partial a_{ij}}$ is given by:

$$\frac{\partial E^*}{\partial a_{ij}} = \left(\sum_{k=1}^{K}\frac{\partial E^*}{\partial z_k}\frac{\partial z_k}{\partial v_k}\frac{\partial v_k}{\partial y_j}\right)\frac{\partial y_j}{\partial u_j}\frac{\partial u_j}{\partial a_{ij}} = \begin{cases} Q_j^*, \mathit{for}\ i = 0 \\ Q_j^* x_i, \mathit{for}\ i = 1,\ldots,I \end{cases}$$

where

$$Q_j^* = \left[\sum_{k=1}^{k} P_k^* b_{jk}\right] y_j(1 - y_j).$$

3.6.3.2 Fuzzy Clustering

K-means and Fuzzy C-means are two conventional clustering methods. The difference of the K-means clustering and the Fuzzy C-means clustering is on the overlapping or not of the boundaries between the clusters. In K-means clustering, the belonging of a datum x to a cluster k or not is crisp, usually denoted by a *membership function* $u_k : X \to \{0,1\}$, where $u_k(x) = 1$ if and only if $x \in k$, and $u_k(x) = 0$ if and only if $x \notin k$. The task of the K-means clustering algorithm is to determine the K cluster centers and the $u_k(x)$ values for every datum and cluster.

In the real life situations, boundaries between the classes may be overlapping (Sarkar et al. 1998) and it is uncertain if a datum belongs completely to a certain cluster.

This is one of the motivations for our adoption of Fuzzy C-means clustering here. In Fuzzy C-means clustering, the membership function u_k is no longer crisp. Instead, here, it can take any values between 0 and 1, with the constraint

$$\sum_{k=1}^{K} u_k(x) = 1$$

for every datum x and every cluster k.

The objective of applying fuzzy clustering component in our study is to minimize the above performance function of the neural network with respect to the w_{knm}, where

$$w_{kmn} \in \{0,1\},\ and \sum_{k=1}^{K} w_{kmn} = 1 \forall m,n$$

Define

$$E_{knm} = \frac{1}{2NM}(z_{knm} - t_{knm})^2$$

as the dissimilarity measure between the object nm and the k cluster center, we can have

$$E = \sum_{k=1}^{K}\sum_{n=1}^{N}\sum_{m=1}^{M} w_{knm}^{\alpha} E_{knm}$$

which can be recognized as a fuzzy clustering problem like (Huang 1998; Huang and Ng 1999), etc., and can be solved by taking partial derivative of E with respect to w_{knm}. For $\alpha > 1$, the minimizer $\hat{w}_{knm}$ is given by

$$\hat{w}_{hnm} = \begin{cases} 1, & if\ E_{hnm} = 0 \\ 0, & if\ E_{hnm} = 0\ for\ some\ other\ k \neq h \\ 1 / \sum_{k=1}^{K} \left[\frac{E_{hnm}}{E_{knm}} \right]^{1/(\alpha-1)}, & otherwise \end{cases}$$

where $1 \le h \le K$ and $1 \le h \le K$.

3.6.3.3 Hybrid Neural Network and Fuzzy Clustering (NN-FC)

Instead of clustering the data only once and then passing it to the neural network, our algorithm further utilize the information of the clustering and neural network. It works in a hybrid iterative loop. As said, the motivation is that outputs from each of the two components can improve each other component in the following round. The algorithm is given as follow:

Algorithm 3.1 *The Hybrid NN-FC Algorithm*

Step 1. Randomize the fuzzy membership matrix w_{knm}.

Step 2. Train the neural network models to minimize E for each cluster, using our modified gradient-descent rules and keeping w_{knm} as constants.

Step 3. Update the fuzzy membership values w_{knm} of every object in each cluster, such that E is minimized with respect to w_{knm}.

Step 4. Repeat the above steps (2) and (3) in iteration until the improvement of the performance function between the successive iterations drops below a certain level.

Then, the accuracies of the clustering results will be checked in our testing period.

3.6.4 Experimental Results with the Simulated Time Series

We have carried out some tests of the NN-FC algorithm on different synthetic datasets that cannot be solved well by the previous studies of NN-FC that are in loose hybrid format (Sarkar et al. 1998; Ronen et al. 1998). And the results show that the proposed algorithm is capable of clustering the data sets accurately and forming the regression functions accurately.

The first dataset comes from two linear generating functions as shown:

$$F_1 : \varepsilon(X(t))$$

$$F_2 : X(t)/10 + \varepsilon(X(t))$$

where $X(t)$ will have values drawn evenly from the interval studied, ε is the corresponding noise. In another example of dataset 2, one generating function is of second-order and the other of third order, with noises at different levels:

$$F_1 : 260 \times (X(t) - 7.5)^2 + 5 + 0.1 \times \varepsilon(X(t))$$

$$F_2 : 8 \times X(t)^3 + 0.1 \times \varepsilon(X(t))$$

The following experiments are with the datasets 3 and 4 respectively. The dataset 3 show us a problem from two linear generating functions with intersection:

$$F_1 : -5.6 \times X(t) + 90 + \varepsilon(X(t))$$

$$F_2 : 6 \times X(t) + \varepsilon(X(t))$$

In the dataset 4, the two generating functions are of second-order and first-order respectively:

$$F_1 : 10 \times (X(t) - 6.5)^2 + 5 + 0.1 \times \varepsilon(x(t))$$

$$F_2 : -68 \times X(t) + 620 + 0.1 \times \varepsilon(X(t))$$

Table 3.1 Clustering accuracies (in percentage) for the datasets in the testing period

Datasets	1 (%)	2 (%)	3 (%)	4 (%)
Our method	96	100	99	97
Sarkar's	61	50	61	77
K-means	62	51	61	77
Hierarchical	51	100	55	63
Quantum	59	50	62	80

Table 3.2 Clustering accuracies (in percentage) for the datasets in the testing period

Noise Levels	0.1	400	1600
Accuracies	96%	91%	96%

And, Table 3.1 shows us the clustering results for these datasets. We can see the accuracies of our proposed method are close to 100% for all datasets in the testing period. Sarkar's results are obtained with fuzzy membership assignment. In their methodology, each of the data sets will pass the fuzzy clustering and neural network components once and only once. While the neural network component has been added as a classifier, the actual clustering jobs are done with the fuzzy clustering. K-means clustering results are also listed. As said, Sarkar's method has employed the fuzzy clustering technique for the clustering process and we can see that their performances are very similar with the K-means clustering. The hierarchical cluster results are obtained with the single linkage algorithm. The quantum clustering algorithm has been recently suggested by Horn and Gottlieb (2002) and has been successfully applied to the genome datasets (Horn and Axel 2003). The algorithm is based on the solution for the Schrödinger equation.

We have applied different noise levels to the first dataset to check the robustness of our method. The following generating functions are that for dataset 1 with different noise levels controlled by α:

$$F_1 : 260 \times (X(t) - 75)^2 + 5 + \alpha \times \varepsilon(X(t))$$

$$F_2 : 8 \times X(t)^3 + \alpha \times \varepsilon(X(t))$$

The noise level 0.1 is the one we have used and is with noise level α equal to 0.1. And the noise level 400 is with α value equal to 400 and the noise level 16 is with α value equal to 1,600. This is comparable with the magnitude of our datasets and we can see output for noise level 4. Table 3.2 shows that the clustering results of these different noise levels. We can see that our method is robust over these levels and can maintain its clustering accuracies above 90% while other methods can give only about 60% accuracy for the original dataset.

Three different neural network structures have also been tested to see the effect of different NN architectures. Among the test structures, they are three-layer networks of five hidden neurons, ten hidden neurons and 15 hidden neurons respectively. All of the three networks with sample 2 datasets can produce stable clusters in less than 20 iterations and obtain 100% clustering accuracies. We can see these results in the below figure. This suggests that our hybrid NN-FC algorithm is stable with respect to the network's structure. And we can observe the total training errors decrease rapidly and converge to a small value in just a few epochs.

3.6.5 Discussions and Further Works

In this illustrative example, it is shown that our NN-FC is able to handle problems that cannot be properly solved with the K-means, fuzzy c-means, and the previous fuzzy clustering and neural network systems. In the studies (Sarkar et al. 1998; Ronen et al. 1998; Boca and Park 1994; Bortolan and Pedrycz 1998), the fuzzy clustering has been designed as a preprocessing tool for feeding the neural network with better inputs without any feedback to the fuzzy clustering. Here, we have successfully developed a feedback algorithm so that the fuzzy clustering can be further improved for solving unsupervised learning problems. And the steady decreases of the sums of errors confirm with this. Table 3.2 shows us that our algorithm can have stable solutions for a range of different noise levels.

There is still much room for the optimization of this methodology. It is expected that further improvement can be obtained with the optimal design of network structure like the number of hidden neurons, with faster second-order learning laws of fuzzy performance function, and with the tabu-search algorithm. Therefore, the proposed algorithm can explore the solution space beyond local optimality in order to aim at finding a global optimal solution of unsupervised learning problems (Ng and Wong 2002).

Chapter 4
Real-Word Application I: Developing Innovative Computing Algorithms for Business Time Series

Abstract Traditionally, business time series forecasting has been dominated by linear methods, which are easy to implement. They are also easy to understand and interpret. However, the business forecasting is a very difficult task, because the processes can behave more like a random walk process and may be time varying. The linear models have serious limitation with problems of nonlinear relationships. It may be unsatisfactory to approximate the linear models for these nonlinear relationships. In business organizations, forecasting is one of the most important activities that form the basis for strategic and operational decision (Zhang 2004). The importance and complexity of the business time series forecasting problem paves way for the importance of innovative computing paradigms. With the progress of the globalization, the effects from other markets may affect markets in other regions. The interdisciplinary innovative computing techniques can be applied to understand, model and design systems for business forecasting.

Keywords Innovation computing • Business time series • Forecasting • Hybrid intelligent models

4.1 Business Time Series

A time series is a set of ordered observations on a quantitative characteristic of an event at time points. When the event concerned is about business phenomenon, it may be called business time series. The frequency of the observations may or may not be equally spaced. A major goal of the business time series analysis is to forecast the future values of the series, i.e., business forecasting. In time series analysis, the past behavior of a time series is used to infer about its future behavior. The causality between the variables is usually unknown. Five key features of economic and business time series are trends, seasonality, somehow influential data points, a changing variance, and non-linearity (Franses 1999). Many methods have been developed for the business time series analysis and forecasting.

S.-I. Ao, *Applied Time Series Analysis and Innovative Computing*,
Lecture Notes in Electrical Engineering 59, DOI 10.1007/978-90-481-8768-3_4,

Methods like linear extrapolation can be used in simple deterministic modeling, while more advance methods like complex stochastic modeling can be used for adaptive forecasting. The time series forecasting can be very helpful for business activities like retailing, manufacturing, and customer services.

4.2 Advances in Business Forecasting

4.2.1 Basic Econometrics Models

Econometrics models are popular in business forecasting. Econometric time series models are constructed for forecasting. The analysis is a major field of the business research and application. Econometric tests have been employed by the financial experts for the analysis of the interdependence between stock markets. The econometric analysis includes methods like linear regression, nonlinear regression, generalized regression, vector autoregression, panel data study, systems of regression equations and regressions with lagged variables, etc. Methods on topics like modeling trends, testing for unit roots, testing for stationarity, seasonality, conditional heteroskedasticity, and nonlinearity, have also been developed. The econometric analysis begins with a set of propositions about some parts of the economy. It is to specify a set of precise, deterministic relationships among variables. These empirical studies can provide estimates of unknown parameters in the model.

4.2.2 Neural Computing Models

Besides the econometric models in business forecasting, there are also nonlinear systems for the business forecasting tasks, like for example the neural network. The neural network is well-known for its capability of pattern recognition and has been playing a more and more active role in the forecasting field. The adopting of neural networks for time series analysis began in the mid-1980s, and then neural networks emerged as an important tool for business forecasting, especially for modeling complex nonlinear relationships. In financial applications, neural networks have been used for predicting exchange rate, interest rate, futures prices, capital market index, property value, and many others. Lapedes and Farber (1987) were among the first to explicitly use the neural networks in this field. They demonstrated that feed-forward neural networks could be useful in modeling deterministic chaos. Since then, there have been more and more researches on the predictability of the neural networks for asset pricing movement. Among these researches, there are the comparisons of the neural networks with the classical regression models.

The results show that in most cases the neural networks are better, or at least are equivalent to classical techniques.

4.2.3 Evolutionary Computing Models

Evolutionary computing has also proved its applicability for time series forecasting in a number of studies. For the forecasting of time series from non-static data-generating processes, the dynamic forecasting genetic program model was proposed by Wagner and Michalewicz (2008). The model was shown to improve upon the performance of genetic programming and other benchmark models for a set of simulated and real time series. Lipinski et al. (2007) proposed an evolutionary approach for portfolio optimization. Three evolutionary algorithms based on evolution strategies are employed to build the risk-optimal portfolio. The performance test was conducted with the financial time series from the Warsaw Stock Exchange. Bautu et al. (2008) proposed a tailored genetic algorithm to find the change points of a time series. Change points in time series occur due to variations in the data generation process. The proposed evolutionary approach employs a new representation and suitable genetic operators. The results showed that the approach can be used for change point detection, and has good potential for time series modeling.

4.2.4 Hybrid Intelligent Models

Due to the complex nature of the business time series, it is sometimes difficult to apply a single intelligent technique to model the characteristics of the time series. Hybrid intelligent models consisting of different components are developed with the objective to combine their respective advantages. In the work of Aburto and Weber (2003), a hybrid forecasting system combining ARIMA models and neural networks was proposed for demand forecast in a supermarket. A replenishment system was developed based on the respective forecasts for the supermarket chain. Wang et al. (2009) proposed a novel hybrid intelligent forecasting model based on empirical mode decomposition (EMD) and support vector regression (SVR). The EMD can adaptively decompose the time series into a finite set of intrinsic mode functions (IMFs) and a residue, with simpler frequency components and higher correlation. The SVR are used to forecast the tendencies of these IMFs and the residue. The model was shown to perform satisfactorily in forecasting of the securities index. Wu et al. (2001) proposed a hybrid neural networks and fuzzy logic system, called FeedForward NeuroFuzzy model, to make the financial forecasting. The model breaks down a large problem into manageable chunks, and achieves better computational efficiency and prediction accuracy. The approach can also improve the transparency problem of the conventional neural networks.

4.3 Developing a Hybrid Intelligent Econometrics Model for Business Forecasting

In global business, the interactions between different markets cause collective lead–lag behavior having special statistical properties which reflect the underlying dynamics. The internal structure of a complex system can manifest itself with correlations among its components. In this section, a cybernetic system of combining the vector autoregression (VAR) and genetic algorithm (GA) with neural network (NN) is outlined how to take advantage of the lead–lag dynamics, to make the neural network forecasting process more transparent and to improve the NN's prediction capability. Two business case studies are described to demonstrate the advantages of our proposed system. The first one is the tourism demand forecasting for the Hong Kong market. Another business case study is the modeling and forecasting of Asian Pacific stock markets. The multivariable time series data is investigated with the VAR analysis, and then the neural network is fed with the relevant variables determined by the VAR analysis for forecasting. Lastly, genetic algorithm is used to cope with the time-dependent nature of the co-relationships among the variables. Experimental results show that our system is more robust and makes more accurate prediction than the benchmark neural network. The novel application of the forecasting modules and the high degree of transparency of the forecasting process will be discussed.

4.3.1 Vector Autoregression

The vector autoregression techniques (Enders 1995; Greene 2000) are used to understand their interactions among different variables. The VAR model is a useful modeling method for business forecasting (Masih and Masih 2001), and is widely used for forecasting macroeconomic time series. Masih and Masih (1999) considered the dynamic causal linkages amongst nine major international stock price indexes over the period 1982–1994. The results showed the significant interdependencies between Asian markets and the leadership of the US markets. The long- and short-term dynamic linkages among international and Asian emerging stock markets were examined with the vector error-correction modeling and level VAR modeling. At the global level, the results showed the leadership of the US over both the short and long term. It also showed the existence of a significant short and long term relationship between the established OECD and the emerging Asian markets. At the regional level in South-east Asia, the results showed the leading role of Hong Kong.

VAR is a multivariate system of equations that we do not need to take the dependence versus independence into account. The basic idea of VAR is to treat all variables symmetrically. Mathematically, a VAR model is expressed as followed (Sims 1980):

$$\overrightarrow{y_t} = \overrightarrow{C} + A(L)\overrightarrow{y_t} + \overrightarrow{e_t}$$

where $\vec{y}$ is a $(n \times 1)$ vector of variables, $A(L)$ is a $(n \times n)$ polynomial matrix in the backshift operator L with lag length p, such that $A_1 L + A_2 L^2 + \ldots + A_p L^p$. $\vec{C}$ is a $(n \times 1)$ vector of constant terms. $\vec{e}$ is a $(n \times 1)$ vector of white-noise error terms. For a model of n variables and n separate equations, the coefficients can be estimated by ordinary least squares. Using the same lag length for all variables produce an equation of the model which have $(n \times p)+1$ coefficients. Statistically insignificant variables and lags can then be excluded in the model. In our experiments, the vector autoregression models are implemented with the RATS, a fast and efficient econometrics and time series analysis software package.

There are researchers like Refenes et al. (1994) who indicated that the conventional statistical techniques for forecasting have reached their limitation in applications with nonlinearities in the data set such as stock indices. Studies like Zhang (2003) proposed a hybrid methodology that combines both ARIMA and ANN models to take advantage of the unique strength of ARIMA and ANN models in linear and nonlinear modeling. The experimental results show that the combined model can be an effective way to improve forecasting accuracy achieved by either of the models separately. Yu et al. (2005) proposed a novel nonlinear ensemble forecasting model integrating generalized linear autoregression with artificial neural networks. Empirical results with the exchange rate data reveal that the prediction with the proposed hybrid nonlinear ensemble model can be improved.

It is suggested that the relationship between neural networks and the traditional statistical approaches for time series forecasting is complementary by White (1989). An advantage of vector autoregression is that multiple variables can be investigated at the same time. Results from the VAR analysis provide the direction of interaction and the quantitative amount of interdependence among specific markets. This characteristic is suitable for our hybrid system to study the interactions among the Asian Pacific markets and the tourist markets without pre-defined assumption.

4.3.2 Neural Network

Neural networks are being applied to a wide variety of tasks in many different business fields. The advantages of neural network include the capability to implicitly detect complex nonlinear relationships and interactions between dependent and independent variables, the requiring of less formal statistical training, and the availability of multiple training algorithms (Smith and Gupta 2000). The users may not necessarily need to predetermine the relationship between inputs and outputs with the exact functional form. Instead, it is decided by the data. Theoretically, it can approximate any functional forms of the input–output pair and can be used in the regression analysis.

The neural network is a backbone of many data mining products available, as the network is known for its non-linear capability. There is a rapid acceptance of the neural networks in the traditional domain of the operations researchers, like forecasting, modeling, clustering, and classification. Neural networks have been successfully applied to business forecasting, a main area of the banking and finance industry, while neural networks have also been applied in various operation planning and control activities like the demand forecasting, shop floor scheduling, and control (Garetti and Taisch 1999). This non-linear method can provide more accurate numerical forecasting.

Business forecasting has been studied with linear statistics for a long time. The traditional approaches to time series prediction, such as the Box–Jenkins or ARIMA method (Box and Jenkins 1976), need to assume that the time series under study are generated from linear process. Linear models have advantages like the ease of explaining the results. They are also relatively easy to be implemented. However, they may be inappropriate if the underlying mechanism is nonlinear. The formulation of a nonlinear statistical model to a particular data set is difficult as there can be too many possible nonlinear patterns. In business forecasting, the neural networks provide an attractive alternative nonlinear tool for both forecasting researchers and practitioners.

Werbos (1974) compared the performance of the neural network trained with back-propagation with the traditional statistical methods such as regression and Box–Jenkins approaches. The neural networks outperformed the other traditional methods. Since then, there were different studies of how to implement neural networks for forecasting. Most studies use the straightforward MLP networks (Kang 1991). Yao and Tan (2000) applied the neural networks in foreign exchange rates for forecasting between American Dollar and five other major currencies, Japanese Yen, Deutsch Mark, British Pound, Swiss Franc and Australian Dollar. Technical indicators and time series data were fed to neural networks to capture the underlying rules of the movement in currency exchange rates. The experimental showed that useful prediction can be made. Swanson and White (1997) applied the NN for predicting future values of nine macroeconomic variables. Abraham et al. (2003) applied an artificial neural network trained using Levenberg–Marquardt algorithm, Support Vector Machine (SVM), Takagi–Sugeno neurofuzzy model and a Difference Boosting Neural Network (DBNN) for the prediction of the Nasdaq-100 index and S&P CNX NIFTY stock index. Experiment results reveal that all the connectionist paradigms considered could represent the stock indices behavior very accurately. The results indicate that the neural network often outperforms a variety of non-adaptive models, linear models like the VAR models, and even professional forecasters as well, when used to predict changes in various macroeconomic variables.

In the typical neural network, there are three layers – the input layer, the hidden layer and the output layer. All these layers are connected and the architecture of the neural network design is itself a worthy field. The inspiring idea for the neural network is to mimic the working of our brain. It consists of axons for inputs, synapses, soma, and axons for outputs. The typical three-layer neural network architecture is employed in our research works. The layers are the input layer, the hidden layer and the output layer. The above layers correspond to the axons for inputs, synapses, soma, and axons for outputs. The mathematical structure for the neural network structure can be expressed as followed (Principe et al. 2000):

$$y = g\left(\sum_{j=1}^{J} w_j^{(2)} f\left(\sum_{i=1}^{I} w_{ji}^{(1)} x_i\right)\right)$$

where I denotes the number of inputs, J the number of hidden neurons, x_i the ith input, $w^{(1)}$ the weights between the input and hidden layers, $w^{(2)}$ the weights between the hidden and output layers.

The training of neural network by back-propagation algorithm has become a standard of operation in business applications, sometimes even a byword for supervised neural networks. To make comparison easier with other studies, the training of neural network in the study also uses the back-propagation algorithm:

Algorithm 4.1 *Back-propagation algorithm of NN*

1. Present the input vector patterns to the network.
2. Propagate the signals forwards, and calculate.

$$u_j = a_{0j} + \sum_{i=1}^{I} a_{ij} x_i, \quad v_k = b_{0k} + \sum_{j=1}^{J} b_{jk} y_j,$$

$$y_j = g\left(u_j\right), j = 1,\ldots,J, \; z_k = g\left(v_k\right), k = 1,\ldots,K$$

3. Calculate the mean square error.

$$E = \frac{\frac{1}{2}\sum_{n=1}^{N}\sum_{k=1}^{K}\left(z_{kn} - t_{kn}\right)^2}{NK}$$

4. Update the weights according to the delta rule.

$$w^{m+1} = wm - \lambda d^m, \; d^m = \sum_{n=1}^{N}\left(\frac{\delta y}{\delta x}\bigg|_m\right)_n$$

5. Repeat the above steps 2, 3, 4 until the error is less than the predefined value or for a predefined number of iterations.

The business applications show that neural networks are a powerful tool on their own. Despite the success of these business applications, the neural network has been criticized for its "black box" nature, proneness to over-fitting, and the empirical nature of model development (Tu 1996). The neural network cannot easily be determined which variables are the most important contributors to a particular output. A neural network model may also contain a number of insignificant predictor variables that the developer fails to figure out. There are not well-established criteria for interpreting the weights in a connection weight matrix. It is an actively developing field to investigate techniques to increase the understanding of the internal logic of neural networks. One proposed technique (Baxt 1992) is to train the neural network with each input variable node removed one at a time and then to observe the effect on network performance. Gnfhth et al. (1992) started to develop regression-like technique to examine the connection weights of various input variables and then determine which variables can be removed from a model without affecting its performance. Nevertheless, none of these techniques has achieved widespread use, as they do not offer the ease of interpretation of the odds ratios associated with the coefficients of a regression model. It will likely hasten their acceptance with the reducing of the black box nature of neural networks.

Synergy can be gained from the integration of neural networks within more general systems. The mixture of the predictions of different neural networks can improve their individual performance (Markham and Ragsdale 1995). The performance of the neural network is expected to benefit from the assistance of other techniques in complex models (Vellido et al. 1999). In other words, they can be integrated into a more general schema of a hybrid system composed of a mixture of models. It is these potential benefits of the synergy that motivated our works of the hybrid neural network system.

4.3.3 Genetic Algorithm

Genetic algorithms were first developed by John Holland in the 1970s (Holland 1975). GAs are parallel, adaptive search algorithms inspired by the mechanisms of biological evolution. The idea of genetic algorithm (GA) is inspired by the concept of natural evolution, which is formulated by Charles Darwin in the nineteenth century. GA can be regarded as a broad collection of stochastic optimization algorithms that let the fittest to survive and the weak to die. The algorithms have been applied to solve a variety of optimization problems (Goldberg 1989). Genetic algorithms were used to find an analytical function that best approximated the time variability of the studied exchange rates (Alvarez-Diaz and Alvarez 2003). In all cases, the mathematical models found by the GA predicted slightly better than the random walk model. Jeong et al. (2002) built a generic forecasting model applicable to supply chain management. A linear causal forecasting model was proposed and its coefficients were efficiently determined using the genetic algorithms. The results showed that it greatly outperformed the regression analysis.

GAs start with a population of solutions to a problem and then attempt to produce new generations of solutions that are better than the previous ones. Dual representation of individuals is needed, one for the 'genotype' (representation space), and another for the 'phenotype' (problem space). A mapping function is also needed between these two representations. In biological systems, a genotype is made up of chromosomes, and the phenotype is the actual organism formed by the interaction of the genotype with its environment. Based on evolutionary theory of natural selection, the chance of producing optimal structures is greatly enhanced by GAs. In GA, the whole solution sets are called the population while an individual solution is referred to as a chromosome. In a chromosome, there exist different characteristics that are represented as the gene. They are corresponding to the different properties of an individual. There exist many generations in GA. The individuals try to reproduce in each generation. The procedure codes of GA are as followed (Tettamanzi and Tomassini 2001):

Algorithm 4.2 *The Procedure Codes of the Genetic Algorithm*

1. Initialize *P(t)*.
2. Evaluate *P(t)*.
3. Recombine *P(t)* to yield *C(t)*.
4. Evaluate *C(t)*.
5. Select *P(t+1)* from *P(t)* and *C(t)*.
6. Repeat the above three steps (3, 4, 5) in the next generation *t+1* until the termination condition is met, where *t* is the order of generation, *P(t)* is the population set at the generation *t*, *C(t)* is the population set after reproduction in the generation *t*.

Genetic algorithms have been employed to improve the performance of neural networks. Williamson (1995) applied the GA to select the optimum topology of a neural network. GA, as a type of unstructured search, can be used to assist neural networks in the task of variable selection (Back et al. 1996). In Hansen et al. (1999), genetic algorithms were used to evolve connection topologies for NNs having a fixed number of hidden layers and a fixed number of computational units in each layer. The performance of the proposed model and autoregressive integrated moving average forecasting models is evaluated within six different time series examples. Refinements to the autoregressive integrated moving average model improve forecasting performance over standard ordinary least squares estimation by 8–13%. In contrast, neural networks achieve dramatic improvements of 10–40%. Goh (2000) combined NNs and GAs to allow for the search of optimum NN structures for construction demand forecasting in Singapore. The combined model outperformed the basis NN model remarkably. Leigh et al. (2002) used a genetic algorithm to determine the subset of input variables to improve the R^2 correlation between the neural network estimated price increase and the actual, experienced price increase.

4.3.4 A Cybernetic Framework of Hybrid Vector Autoregression, Neural Network and Genetic Algorithm

Cybernetics had a crucial influence on the birth of various modern sciences like artificial neural networks, computer modeling and simulation science, and dynamical systems. Cybernetics is a science that studies the abstract principles of organization in complex systems (Heylighen and Joslyn 2001). The focus is on how systems use information, models, and control actions to steer towards and maintain their goals amongst various disturbances. This goal-directed behavior can be done with negative feedback control loops which try to achieve and maintain goal states. Cybernetic systems can provide an ideal comparison for many of the dynamics at work within this form of adaptive cycle. A simple example of such a system can be a domestic central heating system, composed of an appliance and a regulator. The regulator can adjust the heating level in according with the detections of changes in temperature inside the room. This negative feedback can achieve its predetermined goal.

A cybernetic system is a system that can dynamically match acquired information to selected actions relative to a computational issue that defines the essential purpose of the system or machine (Fry 2002). It is necessary that the information and control need to be quantified. In term of automation, a simple cybernetic model can be regarded as consisting of automata, a system that transforming one set of signals into another one. Each automation receives one or more input signals which it transforms into one or more output signals. An automation represents a simple function operating on input signal or signals, while the number of input signals received and output signals sent may vary among automata. In business domain, Morgan and Hunt (2002) discussed with a conceptual and theoretical view about cybernetic systems of scenario planning in marketing strategy.

Modular hybrid systems refer to those systems that are modular in nature, i.e. they are comprised of several modules which can have different degrees of coupling and integration. An important feature is that they do not involve any changes regarding the conceptual operation of the individual modules (McGarry et al. 1999). A vast majority of hybrid systems fall into this category. The main reason is that these systems are powerful processors of information and are relatively easy to implement. Our works of the hybrid vector autoregression, neural network and genetic algorithm (VAR–NN–GA) framework can supplement its separate standalone components. The framework of the VAR–NN–GA is as followed:

Algorithm 4.3 *Framework of the proposed VAR–NN–GA system*

Input: Multivariable Time Series Data (MTS) like index, visitor number, GNP, etc.
Implementation Procedure:

1. Pass the MTS data through VAR module.
2. Test the variables against each other to see their respective significance levels.
3. Select the variables and lag terms that are within the confidence interval.
4. Formulate the input vectors for the neural network from the above selected MTS.

5. Make one-step ahead forecasting with the NN models for the first out-of-sample time point.
6. Apply the GA module to assign dynamic weights to the NN models based on the prediction performance of the time point in step 5.

Steps 5 and 6 will then repeat with the forecasting for next out-of-sample time point, and so on, until finishing all out-of-sample time point.

Previous studies have recognized the close relationships between the economic learning models and models of evolutionary computation. The objective of our framework is to develop a system that can take over the task of the model selection process. The system has the capability to learn which model should be employed for different sub-periods. The goal of the GA is to let the strategy with higher payoff to remain while the ones with lower payoff more likely to disappear. This methodology is suitable for our evolutionary purpose. We can image that there exist many prediction models. The evolutionary process comes into play to let the model with better prediction to survive more likely than the ones with poor records. The objective function is defined to minimize the sum of absolute errors of the predictions.

The methodology can work with inputs data like indices of the regional countries. Input data, like the trade volume, economic growth rate and currency exchange rate, etc., can also be tested by the VAR analysis. For the input variables with the lowest significance level, they will be used as the input variables for the neural network. The performance of the predictions made by the neural network with these input variables is time-dependent and unstable. In some sub-periods, some input variables may be fitter for the prediction, while, in other sub-periods, they may be poorer in fact. These neural network prediction models with different inputs can be regarded as experts of different opinions of the relevant input factors. Their predictions may vary with time, as said. The selection and evaluation of these predictors can be made in the evolutionary cycle. Experts with higher forecasting accuracy in each cycle are going to weigh more heavily in the coming round.

In the following two business applications, we will illustrate the details of how to implement this cybernetic approach to quantify the cross-market dynamics and to make accurate forecasting.

4.4 Application for Tourism Demand Forecasting

In our first case study, the most recent historical value of the tourist arrival number is used for the prediction, serving as the feeding data for the neural network. The forecasting of tourism demand is important for various reasons. The hotel booking, the preparing for the food and transportation, and the airline seat-reservation are a few examples that require accurate prediction. Relatively recently, neural network is introduced into the tourist forecasting and it is found to be superior to other methods (Law and Au 1999).

It is important to make accurate forecasting of tourist numbers for various reasons. Hotels, restaurants, ground transportation companies, and the airline corporations are a few examples that require forecasting results to be as accurate as possible. Because of the perishable nature of the tourism industry, the need for accurate forecasts is crucial. It is not possible to stock the unfilled airline seats, unoccupied hotel rooms, or unused facilities. Forecasting of tourism demand provides a guide to the efficient allocation of resources for the tourism industry. The tourism demand is usually measured by the number of tourist arrivals. The analysis is an integral part of sound decision making regarding investments in both the public and private sectors. It is also important for the short-term marketing decisions to promote tourism products and services. Nevertheless, there is no standard supplier of tourism forecasts (Witt and Witt 1995).

4.4.1 Quantifying Cross-Market Dynamics

In this case study, the vector autoregression (VAR) models were employed to test the interdependence of tourist numbers from different countries to Hong Kong. The monthly and quarterly tourist data of the number of visitors to Hong Kong were used. The data is from January 1978 to December 2002, available from the DataStream. Separate origins of the visitors are available for those from Japan, South East Asia, West Europe, USA, Australia and New Zealand, Canada, and others, denoted by HKVISJAF, HKVISASIF, HKVISWEF, HKVISUSA, HKVISANZF and HKVISCANF respectively. The total arrival number is denoted by HKARRIVL. The GDP and population data are from the International Financial Statistics.

The tourism demand time series data of various source countries are tested with its own lagged terms as well as against the time series of population and GDP. Our research results have shown that the patterns of these time series data are not clear. Even though some lead–lag relationship is statistically significant, its importance on the forecasting of future visitor number is not as useful as the time series data of that country itself. This is mainly in agreement with the study (Law and Au 1999). The hypothesis is that the variables between markets are independent. The significance level tells us the probability that the hypothesis is found invalid. For example, with significance level at 0.022 for the second lag term of HKVISJAPF, denoted by HKVISJAPF{2}, on HKVISJAPF, it means that there is only statistically 2.8% of the independence of these two variables. In the order words, they are likely to be correlated.

4.4.2 Experimental Results

Uysal and Roubi (1999) applied the NN in tourism demand studies. The study used Canadian tourism expenditures in the United States as a measure of demand to demonstrate its application. The results revealed that the use of NNs in tourism demand studies may result in better estimates in terms of prediction bias and accuracy than the multiple regression. Multiple regression using ordinary least squares (OLS) has been the most widely used approach in international tourism demand analysis. Palmer et al. (2006) designed a neural network for tourism time series forecasting. The time series corresponding to tourism expenditure in the Balearic Islands (Spain), one of the world's major tourist destination, has been used. The experimental results support that the NN model can be applied successfully to tourism data forecasting. Law and Au (1999) proposed the application of a supervised feed-forward neural network model to forecast Japanese tourist arrivals in Hong Kong. The experimental results showed that applying the neural network model to forecast Japanese arrivals outperforms multiple regression, naive, moving average, and exponent smoothing.

The hybrid model of support vector machines and genetic algorithms was employed for accurately forecasting arrivals in Barbados (Pai and Hong 2005). The experimental results showed that the proposed models outperformed the ARIMA approaches. Chen and Wang (2007) proposed a novel neural network technique, support vector regression (SVR), to tourism demand forecasting. To build an effective SVR model, SVR's parameters must be set carefully. The genetic algorithm is used to search for SVR's optimal parameters. The tourist arrivals to China during 1985–2001 were employed as the data set, and the experimental results are promising.

In our research works, the relationships between the current visitor number and its previous numbers were studied. The multivariate regression techniques are suitable to assist the understanding of their interactions with each others. The time series of the visitor number data is auto-regressed against its own lagged terms as well as against the time series of population and GDP. The VAR analysis assists us to identify lead–lag dynamics among the variables. Our framework are the first attempt to combine the linear autoregression models with the non-linear hybrid NN and GA modules for the tourism demand forecasting with satisfactory performance.

4.5 Application for Cross-Market Financial Forecasting

The relationship between the US and the Asian markets has been studied. Briefly, in the study (He 2001), the monthly data between US's stock and interest rate and the HK's and SK's stock market was studied using OLS. In the study (Dekker et al. 2001), the US stock market and Asian markets were investigated with daily data using VAR. Soydemir (2002) determined the relationship between US's stock market and macroeconomic factors like the interest rate and the stock markets in US and Latin America. The weekly data before the Mexico financial crisis in 1994 was used. Positive relationship was found for the stock markets while negative relationship of various degrees has been found between the US's interest rate and Latin American's stock market was observed. Our works of the VAR models have also confirmed that the correlation between different stock markets is statistically significant.

4.5.1 *Quantifying the Cybernetic Lead–Lag Dynamics across Different Markets*

The stock prediction analysis derives the future stock movement from its historical movement, basing on the assumption that there exists strong enough correlation for prediction. The historical data can be used directly to form the support level and the resistance or they can be plugged into many technical indicators for further investigation. With the flow of new information into asset markets, the market prices for the assets concerned readjust to such news flows. If two markets do not react at the same time, one market will then lead the other (Bose 2007). If this lead–lag relation happens, the leading market can be viewed as contributing a price discovery function for the lagging market, and this motivate the studies of the cross-market financial forecasting studies. As we have seen, there have been reports of the network's superiority on predictions. Most of these studies use the historical data of one market only, and the neural network uses the time series data for training the network to make prediction. In the stock markets, there exist many different prediction models basing on various opinions and assumptions. For example, one may say that the market in Hong Kong is influenced heavily by its own previous movements, while others may argue that US may also have a strong influence. The studies like (He 2001; Ao 2003a) supports the later one. There is strong correlation between the US market and the Asian markets in the long run. The VAR analysis shows that the US indices lead the Asian indices. But, in our works (Ao 2003a), such correlation is found to be time-dependent/unstable, which affects the performance of using the historical US data to predict the Asian markets by neural network.

In this case study, other Asian Pacific markets' historical data are incorporated into the neural network prediction models. In our vector autoregression analysis, it is shown that the lead–lag relationship does exist between the markets. Then, the neural network is employed to incorporate other leading markets' information into the prediction of the lagging markets. The result has been found to be positive in general. But, it is also found that such correlation is unstable and time-dependent. Additional information of the current correlation magnitude is needed to determine which model is to be adopted. In our studies, a framework with the neural network working along with the genetic algorithm was built to see if such a process can be automated. Our results have shown that the inclusion of highly-correlated market movements can assist the forecasting process. Our works are initial attempts to apply non-linear hybrid system for the forecasting of Asian Pacific stock markets as a whole.

4.5.2 *Benchmark Stand-Alone Neural Network*

There is a well-established tradition in forecasting research of comparing techniques on the basis of empirical results (Adya and Collopy 1998). In a survey of 22 studies that effectively implemented the NN models for business forecasting, the NNs models outperformed alternative approaches in 19 (86%) of these studies.

In our works, the neural network was an integral part of our knowledge discovery and forecasting process and was central to the proposed methodology. The out-of-sample performance of a standard NN model was used as a benchmark to gauge the performance of the proposed cybernetic system.

4.5.3 Hybrid Innovative System and Results Comparison

Studies like Jo et al. (1997) have stressed the lack of formal techniques for NNs to assess the relative relevance among the independent variables. Neural networks are also well-known for its adduced incapacity to identify the relevance of independent variables in business forecasting. There are criticisms that, as the relationship among the variables is not known in advance, the neural network acts just like a black box. This is a motivation for the proposed hybrid system of neural network, econometrics and genetic algorithm. In our works, an automated system is exemplified to overcome this difficulty. With the VAR module, it can be assured that all of our feeding variables for the neural network have significant influences on the target variables. The GA module can enable the system to select the fittest models among the different models. Business markets are dynamics and their properties may be time-varying during different sub-periods. It is desirable to employ the most suitable model for each sub-period. This task may be achieved by the evolutionary computation. The evolutionary computation, including the genetic algorithm, can cope with the market dynamics. Combing these three procedures together, the hybrid VAR–NN–GA system is designed to automate the process of the selection of input variables, the numerical predictions and the evaluations of various prediction models. The aim is to supplement the previous studies, with additional information from other leading markets' movements.

Generally speaking, our cybernetic framework can work with multivariable time series data like indices of the regional countries and other business indicators. Input data, such as the trading volume, economic growth rate and currency exchange rate, etc., can also be tested by the VAR analysis. The input variables with the lowest significance level are used as the input variables for the neural network. In many cases, the performance of the neural network is time-dependent and unstable. In some sub-periods, some input variables may be more suitable for the prediction, while, in other sub-periods, they may be poorer. Neural networks with different inputs can be regarded as experts with different opinions of the relevant input factors. Their prediction performance may vary with time, as said. The selection and evaluation of these predictors can be made in the evolutionary cycle. Experts with higher forecasting accuracy in each cycle are going to weigh more heavily in the coming round. The outline of our hybrid system's modules is as followed:

1. VAR analysis, which is to search for the correlated and leading indicators automatically
2. Neural network prediction, which is to make forecasting from the relevant inputs determined by the VAR analysis

3. Genetic algorithm, which is to cope with the time-dependent nature of the co-relationships among the variables and to adjust the weightings of each neural network model

In our research works, in order to check the performance of our proposed framework, comparison of the business forecasts has been conducted, based on out-of-sample performance with the testing sample different from the training sample. This practice matches with the conditions in real-world business forecasting, with the needs for predictions about an unknown future. The results of our research works were shown to be satisfactory.

4.6 Discussions and Further Works

In the case studies, a hybrid system is illustrated how to automate the process of selecting input variables, to make the neural network module more transparent and to improve the prediction performance with dynamical cross-market lead–lag interactions. For the tourist demand forecasting, different models have been developed separately for making prediction. There are the linear systems like econometric techniques. There are also the nonlinear systems like neural network, which has been known for its capability of pattern recognition and has played a more and more active role in the forecasting field. Our works are the first attempt to combine the linear autoregression models with the non-linear hybrid NN and GA modules for the tourism demand forecasting. Results show that the hybrid system is robust to the changes of the cross-market dynamics.

In the financial forecasting, previous studies have focused more or less on the historical prices and the trading volume of one market only, with either the linear statistical or the nonlinear machine learning approach. Our cybernetic system combines the advantages of both the econometrics, which can offer clear explanation and significant testing for the correlations among the variables, and the machine learning modules NN and GA, which can make our system adaptive to the changing dynamics of cross-market dependences. This is a first attempt of applying this hybrid cybernetic system for the forecasting of Asian Pacific stock markets as a whole. The results show that the performance of the proposed system can be as much as 35% better than that of the benchmark neural network model (Ao 2003b, c, d, e, 2006).

Chapter 5
Real-Word Application II: Developing Innovative Computing Algorithms for Biological Time Series

Abstract In this chapter, recurrent Elman neural networks and support vector machines have been outlined for temporal modeling of microarray continuous time series data sets. An ensemble of the ENN and SVM models are described for how to further improve the prediction accuracy of the individual models. In order to provide the neural networks with explanation capabilities, a pedagogical rule extraction technique is considered for inferring the output of our proposed ensemble system. The outlined pedagogical rule extraction technique is a two-step test of causality and Pearson correlation for the network inference between the causal gene expression inputs and their predicted outputs. The gene regulatory network can be reconstructed satisfactorily with this hybrid methodology.

Keywords Elman neural network • Support vector machine • Ensemble • Microarray • Time series • Rule extraction • Gene regulatory network

5.1 Biological Time Series

The temporal patterns exhibited by biological time series are often complex, nonlinear or chaotic, nevertheless, the analysis are emerging as important means for many biological studies. To address the nonlinear properties of the biological time series, nonlinear methods become important tools to study these systems. Traditional nonlinear analysis is a powerful approach to understand the biological time series, but, the analysis usually needs very long data sets, which may be difficult or very expensive to obtain in the biological experiments and observations. Varanini et al. (1993) proposed a method for artifact detection based on linear modeling of biological time series. Each time series was modeled by means of adaptive prediction filter. The method was applied to sequences of cardiovascular measurements from ICU and from Holter monitoring with satisfactory results. Richman and Moorman (2000) proposed a new complexity measure, sample entropy to measure the similarity of two distinct time series. The experimental results were satisfactory and the new statistics is useful in the study of not only experimental clinical

S.-I. Ao, *Applied Time Series Analysis and Innovative Computing*,
Lecture Notes in Electrical Engineering 59, DOI 10.1007/978-90-481-8768-3_5,

cardiovascular but also other biological time series. Korenberg (1989) talked about methods for obtaining a parsimonious sinusoidal series representation of biological time series data. The methods could identify nonlinear systems with unknown structure, resolve the sinusoidal signals, and produce an economical sinusoidal series presentation, with the most significant frequencies first.

5.2 Advances in Experimental Designs for Microarray Time Series

5.2.1 Microarray Experiments

Microarray is a solid substrate where the DNA is attached to in an ordered manner at high density (Geschwind and Gregg 2002). Among the high-throughput methods of gene expression, the microarray has been the most widely used one for assessing the differences in mRNA abundance in the biological samples. With the work of Patrick Brown and his colleagues (DeRisi et al. 1996), microarray has been gaining its popularity.

In a single microarray experiment, the expression levels of as many as thousands of genes can be measured simultaneously. Thus, it can enable the genome-wide measurement of gene expression. This is a large improvement over the situation of "one gene per experiment" in the past. As a result, microarray has been found useful for different types of biological researches, for example, tissue-specific gene expression, developmental genetics, genetic diseases, complex diseases, and environmental monitoring, etc.

A typical microarray experiment consists of the following five steps (Amaratunga and cabrera 2004):

(a) Preparation of the microarray: Drops of purified single-stranded DNAs is placed onto glass microscope slide.
(b) Preparation of the labeled sample: mRNAs are purified from the sample contents, and then reverse-transcribed into more stable cDNA or cRNA.
(c) Hybridizing of the labeled sample: Label sample is then sealed in hybridization chamber for hybridization reactions.
(d) Scanning of the microarray: This is to check the amount of labeled sample bound to each spot of the microarray.
(e) Data analysis of the scanned image: The scanning product of the microarray is a gray scale image, and image-processing tools are needed to convert the image into spot intensity measurements for further data analysis.

Multi-channels cDNA microarray and oligonucleotide array (pioneered by Affymetrix) are two popular microarrays. New microarray technologies are also emerging, for example, the bead-based microarray technology.

5.2.2 Microarray Time Series and Applications

The microarray technology can also enable us to have the gene expression values at different time points of a cell cycle. In the simplest case of time series expression analysis, two time points are taken: before and after an event. A more comprehensive study will involve the taking of values at different periods. The frequencies of the time points can have ranges from several minutes to several hours. Various methods like self-organizing maps (Nikkilä et al. 2002), *k*-nearest neighbor (Acta 2001) and hidden Markov models (Ji et al. 2003) have been employed for the microarray analysis. These studies mainly focus on the clustering and the measurement of the similarity among the different expressions. For the gene expression time series analysis, methods like warping algorithms (Aach and Church 2001), the comparison of similarity functions of the genes (Butte et al. 2001), the identification of gene regulatory networks with graph method (Chen et al. 2001), and dynamic models (Dewey 2002), etc., have been developed.

In the literature, different methods have been developed to analyze gene expression time series data, see for instance (Yoshioka and Ishii 2002; Tabus and Astola 2003; Syeda-Mahmood 2003; Wu et al. 2003). The construction of genetic network from gene expression time series is tackled in (Kesseli et al. 2004; Tabus et al. 2004; Sakamoto and Iba 2001). The visualizing of the gene expression time series is discussed in studies (Zhang et al. 2003; Craig et al. 2002). Some special clustering algorithms have been employed to explore the gene expression time series data from the microarray experiments. Costa et al. (2002) have proposed the symbolical description of multiple gene expression time series. Each variable will take as a set of values in a time series and the results are compared with Self-Organizing Map algorithm. Yoshioka and Ishii (2002) have employed a clustering method based on mixture of constrained PCA. It can classify genes with similar expression patterns into the same cluster regardless of their magnitude (scale). In the study, Tabus and Astola (2003) have handled the problem of the non-uniformly sampling of the gene expression time series. The minimum description length model is fitted to each gene and then the optimum parameters are used for clustering the genes. The extrapolation of the gene expression time series data by the minimum description length model can be applied in our methodology too for non-uniformly sampling data.

Syeda-Mahmood (2003) has studied a clustering algorithm that uses the scale-space distance as a similarity metric. The scale-space analysis is to detect the sharp twists and turns of the gene time series and to form the similarity measure between time profiles. Wu et al. (2003) have developed a procedure for the determination of the minimal number of samples or trials required in a microarray experiment for clustering. The procedure is an incremental process that will terminate when the evaluation of the results of two consecutive experiments of k-means clustering shows they are sufficiently close. Jiang et al. (2003) use a density-based approach to identify the clusters such that the clustering results are of high quality and robustness. Futschik and Kasabov (2002) employ the fuzzy c-mean (FCM) clustering to achieve a robust analysis of gene expression time series. The issues of parameter selection and cluster validity are also addressed.

The visualizing of the gene expression time series is discussed in studies (Zhang et al. 2003; Craig et al. 2002). Zhang et al. have introduced the first Fourier harmonic projection (FFHP) to translate the multi-dimensional time series data into a two-dimensional scatter plot. The spatial relationship of the points reflects the structure of the original dataset and the relationships among clusters become two-dimensional. Craig et al. propose the display technique that operates over a continuous temporal subset of the time series, with direct manipulation of the parameters defining the subset. Its advantage is that the number of elements being displayed will not be reduced.

5.3 Reverse Engineering of Biological Networks

5.3.1 Introduction

In the past decade, the advances on high-throughput DNA microarray technologies have enabled the simultaneous measurement of hundreds to thousands of gene expression levels. The construction of genetic network from gene expression time series is tackled in Kesseli et al. (2004); Tabus et al. (2004); Sakamoto and Iba (2001). Kesseli et al. have employed monotonic time transformations (MTT) for inferring a Boolean network. Several different methods of clustering have been used to form different transformations. Tabus et al. build systems of differential equations for specifying the genetic networks. The structure of the networks is inferred by operating with the exact solutions of the linear differential equations, which are obtained through the eigenvalue decomposition of the system matrix. Sakamoto and Iba also use a system of ordinary differential equations as a model of the network and infer their right-hand sides by using genetic programming (GP) instead. The least mean square (LMS) method is used along with the GP to explore the search space more effectively in the course of evolution. In these systems of linear differential equations, there is a strong assumption that the genetic interactions are linear. Instead, with our PCA-NN algorithm, we can have the advantage of the nonlinear flexibility of the neural network. Also, we have employed the AIC test to decide the optimal lag length used in our models, whereas, in the above models, only one lag length of each gene expression value change is included. The lag length refers to the number of lags (the number of previous values of the variable) used in the model.

Langmead et al. (2002) formulate the task of estimating an expression profile's periodicity and phase as a simultaneous bicriterion optimization problem. The maximum entropy-based analysis technique is employed for extracting and charactering rhythmic expression profiles, and is found to work better than the Fourier-based spectral analysis for signals in the microarray experiments.

Yeang and Jaakkola (2003) explain time correlations between gene expression profiles through factor-gene binding information to estimate latencies for transcription activation. This can estimate latencies for transcription activation. The resulting aligned expression profiles are subsequently clustered and again combined with binding information to determine groups or subgroups of co-regulated genes.

Systems Biology studies on gene interactions have gained its prominence for reconstructing gene regulatory networks using the microarray technology (Cho et al. 2007). These studies on gene regulatory networks (GRNs) employ regulation matrix methods, Boolean methods, Bayesian methods and machine learning methods etc. Among the machine learning methods, neuro-fuzzy techniques have been applied to microarray gene expression analysis in recent studies. Azuaje (2001) applied a simplified fuzzy ARTMAP to identify normal and diffuse large B-cell lymphoma patients and to provide interpretation for the genome expression patterns. Neagu and Palade (2003) have applied a neuro-fuzzy system for the functional analysis of gene expression data from microarray experiments. Chen et al. (2006) have applied neuro-fuzzy methods for the identification of critical genes in microarray experiments. The weights of the trained neural network were fed into a fuzzy logic inference system. The approach has been shown to be able to uncover the unique features of host response to individual bacterium. Maraziotis et al. (2007) have proposed a neural fuzzy recurrent network for the time series prediction from microarray data.

5.3.2 *Materials and Methods*

5.3.2.1 Elman Neural Networks

The Elman network is a recurrent neural network of simple architecture which can be trained with the backpropagation (BP) algorithm. Recurrent neural networks have feedback links and incorporate temporal aspects better than feedforward neural networks (Kodogiannis and Lolis 2002). They also have state variables for the delays, so less information is required when modeling a system. The Elman network consists of several layers, with the nodes using the input function, the weight function and the transfer functions (activation functions). The first layer is the input layer and the inputs are weighted with the weight function. Besides the output layer, all the other layers have recurrent links (Zhang et al. 2007).

The state layer of the Elman network makes it different from the multilayer perceptron neural network. The state units of the Elman network can memorize all the feed inputs such that the outputs of the network depend upon the current input as well as the previous inputs. With memory and recurrent feedback connections, an Elman network can learn temporal patterns effectively. At each time step, the values of the hidden layer units are copied to the state layer, and this information can be stored for future use. The Elman network has this dynamic memory property which makes it suitable for time series analysis. The network can be expressed by a state space model. In the state space model, the time series is modeled as a linear transformation of a time-dependent state (Dorffner 1996):

$$\vec{x}(t) = \mathrm{C}\vec{s}(t) + \vec{\varepsilon}(t)$$

where $\vec{x}$ is a function of the time variable *t*, $\vec{s}$ is a state vector, *C* is a transformation matrix. $\vec{\varepsilon}(t)$ are the noises. This time-dependent state vector can be modeled as

$$\vec{s}(t) = \mathrm{A}\vec{s}(t-1) + \mathrm{B}\vec{\eta}(t)$$

where *A* and *B* are the coefficient matrices. $\vec{\eta}(t)$ are the noises. This simple model is basically an ARMA[1,1] process. When the states are dependent on the past sequence vector, the state vector can be expressed as (neglecting the above noise $\vec{\eta}(t)$):

$$\vec{s}(t) = \mathrm{A}\vec{s}(t-1) + \mathrm{D}\vec{x}(t-1)$$

where D is a coefficient matrix. The above models are linear state space models. In the Elman network, we have activation functions in the hidden layer with the state vector:

$$\vec{s}(t) = \sigma\left(\mathrm{A}\vec{s}(t-1) + \mathrm{D}\vec{x}(t-1)\right)$$

where $\sigma(\cdot)$ represents the application of the activation function. For example, with the logistic activation function, $\sigma(\vec{a})$ is expressed as: $1/(1+\exp(-a_i))$ for each element a_i of $\vec{a}$. It can be observed that this transformation is no longer linear after the application of the activation functions. Figure 5.1 represents the Elman network with the state-space model.

The individual Elman neural networks can be fed with the gene expression levels individually. In our experiments, the Elman neural network is from the Neural Network Toolbox of the MATLAB 7.0. All our algorithms have been run in a PC

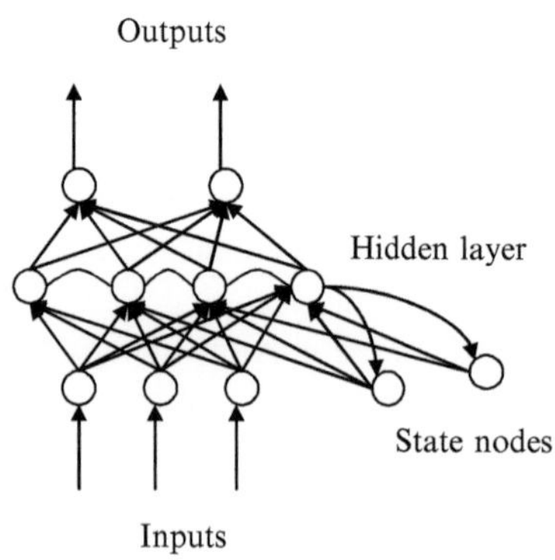

Fig. 5.1 High-level architecture of the Elman neural network, with the two neurons on the *right hand side* as the state nodes, and the input neurons at the *bottom*, while the output neurons at the *top*

of Intel Core Duo CPU 1.73GHz and 1.75 GB of RAM. Then, in the second step, an ensemble of the neural networks of different gene expressions was constructed. As the ensemble is a combination of the different gene expressions and network models, our individual Elman neural network models were fed with input variables one by one.

5.3.2.2 Support Vector Machines

Support vector machines can be divided into two main types: support vector machines for classification (SVC) and support vector machines for regression, in short, support vector regression (SVR). Support vector machines (SVMs) were developed at AT&T Bell Laboratories by Vapnik and his co-workers (Boser et al. 1992). Because of its industrial context, the support vector approach had a sound orientation towards real-world applications (Smola and Scholkopf 2004). Support vector machines have many mathematical features attractive for gene expression analysis (Brown et al. 2000). These include the sparseness of solution for large data sets, the flexible capability for handling large feature spaces, the robust identification of outliers, etc. SVC has found applications in different bioinformatics domains, for example, protein secondary structure prediction (Guo et al. 2004), cancer classification (Guyon et al. 2002), and enzyme family classification (Cai et al. 2004). While SVR has been applied to the time series prediction applications and excellent performances were obtained (Muller et al. 1997) and (Mattera and Haykin 1999), only until recently, the SVR has been applied for the bioinformatics problems like the missing value estimation (Wang et al. 2006a) and the smoothing of expression profiles (Martin et al. 2007) for DNA microarray gene expression data.

Support vector machines are based on the structural risk minimization principle from statistical learning theory. The structural risk minimization principle can minimize both empirical risk and confidence principle and avoid the "over-fitting" problem. The basic principle of the support vector machines is to map the samples from the low-dimension input space into a much higher dimensional space with a kernel function. Then, quadratic programming is applied to search the global optimal solution to the corresponding problem. Support vector regression constructs the fitted regression function by solving this optimal problem with constraints. The term support vectors refers to the samples with non-zero Lagrange multiplier. The support vectors are the input samples that will be used in the regression analysis, and they determine the SVR function. In other words, the non support vectors in the data set will not have any influenced in the regression model. Usually, the number of support vectors is small relative to the total number of samples.

Mathematically, the constrained quadratic programming problem of SVR can be expressed as (Wang et al. 2006):

$$\min_{w,b,\xi,\xi^*} \frac{1}{2} W^T W + C\sum_{i=1}^{n} \xi_i + C\sum_{i=1}^{n} \xi_i^*$$

subject to

$$
\begin{aligned}
&W^T\phi(x_i)+b-z_i \le \varepsilon+\xi_i \\
&z_i - W^T\phi(x_i)-b \le \varepsilon+\xi_i^* \\
&\xi_i,\xi_i^* \ge 0, i=1,\ldots,n
\end{aligned}
$$

where W is the solution of the primal problem and C is the parameter of regulation that makes the tradeoff between margin ξ_i and prediction error ξ_i^*. $\{(x_1,z_1),\ldots,(x_n,z_n)\}$ denotes a set of the training data, with the input $x_i \in R^n$ and the output $z_i \in R^l$. $\phi(x_i)$ is a non-linear function that maps the input feature into the higher dimensional space and ε is the error probability which limit the deviation between the regression function from the target. For a smaller value of ε, more support vectors are needed.

The dual problem for this quadratic programming problem is:

$$
\min_{\alpha,\alpha^*} \frac{1}{2}(\alpha-\alpha^*)^T Q(\alpha-\alpha^*)+\varepsilon\sum_{i=1}^{l}(\alpha_i+\alpha_i^*)+\sum_{i=1}^{l} z_i(\alpha_i-\alpha_i^*)
$$

Subject to

$$
\sum_{i=1}^{l}(\alpha_i-\alpha_i^*)=0, 0\le\alpha_i,\alpha_i^*\le C, i=1,\ldots,l
$$

where $Q_{ij}=K(x_i,x_j)=\phi(x_i)^T\phi(x_j)$. The SVR regression function can be obtained as

$$
\phi(x)=\sum_{i=1}^{l}(-\alpha_i+\alpha_i^*)K(x_i,x)+b
$$

where $K(x_i,x)$ is the kernel function. The required support vectors are the input vectors which have their corresponding α of non-zero value. It should be noted that this kernel function can assume different forms, such as linear function, polynomial function, radial basis kernel function, etc. The performance of the SVR depends on the setting of the kernel functions as well as their corresponding parameters.

The support vector regression algorithm has been applied in the experiment for the microarray time series data (Joachims 1999). The algorithm has implemented a fast optimization algorithm of the following features: working set selection based on steepest feasible descent, "shrinking" heuristic, caching of kernel evaluations, and use of folding in the linear case. It has been shown that the algorithm can handle problems with thousands of support vectors efficiently. In our study, the SVR has been fed with the microrarray datasets, following the same procedure for our neural network models, that is, the input time series is fed one by one and the one-step-ahead prediction is made. The SVM Light V6.01 has been employed. It is an implementation of Support Vector Machines in C, and we have called the SVM Light from the MATLAB environment with a shell script.

5.3.2.3 Ensemble of Innovative Models

Ensemble methods can be applied to numerical prediction and classification problems. Ensemble methods have been developed for improving the predictive performance of a given individual statistical learning algorithm. Previous studies have shown that ensemble methods can often provide more accurate prediction than any of the individual classifiers in the ensemble (Opitz and Maclin 1999). The resulting classifier of the ensemble is generally more accurate than any of the individual classifiers making up the ensemble. Both the theoretical and empirical research has shown that the individual classifiers of a good ensemble need themselves to be accurate and to make their errors on different parts of the input space. Studies like Maqsood et al. (2004) show that ensemble methods can improve numerical prediction above that of the individual predictors.

Here, Elman neural networks and support vector regression methods have been described for the time series prediction with continuous numerical microarray inputs. The results obtained with these two individual models are comparable with the neural fuzzy recurrent network (NFRN) model. For reference, the performance of the adaptive neuro-fuzzy inference system (ANFIS) has also been investigated for the time series microarray data. To improve the time series prediction accuracies, the ensemble method has used Elman networks and support vector machines for temporal modeling. In our works, the prediction results on the testing microarray data sets support that the ensemble method outperformed the prediction of its individual predictors. The hybrid ensemble method surpasses other computational techniques and outperforms the NFRN.

Neural networks are well-known for their universal function approximation capability, but neural networks have also been criticized for their low explanation capability and their lack of transparency in their solutions. Considerable efforts have been made on addressing the tasks of augmenting a neural network with explanation facilities, for example, Tickle et al. (1998). As a recurrent neural network (RNN) processes information with its internal continuous state space as a memory of the past input patterns (Elman 1990), rule extraction techniques have been developed to approximate this continuous state space by a finite set of states. The dynamics of a RNN can be mapped to transitions among this discrete set of states. The rule extraction techniques can mimic the network to a satisfactory degree while they offer the advantage of more transparency (Jacobsson 2005).

The ensemble method has been developed to improve the prediction performance over that of individual predictors. In an ensemble of neural networks, a collection of a finite number of neural networks is applied simultaneously for the same task. This ensemble approach originated from the work (Hansen and Salamon 1990), which have applied many neural networks and combined their predictions to form the ensemble. The results showed that significant improvement can be obtained for the generalization ability of a neural network system. The ensemble neural networks with combinational feature selection have been applied to the microarray experiments for the tumor classification and remarkably improved results have been obtained (Liu et al. 2004). It has been shown that the ensemble method is able to reduce the instability of the individual neural networks and help alleviate the problem of trapping the neural network into local minima. The ensemble method can cancel the noise part among its individual networks and retain the fitting to the regularities of the data.

A basic approach of the ensemble methods is to formulate a linear combination of some individual learning algorithms, instead of using one single fit of one algorithm. An estimation of a real-value function can be expressed mathematically as $g:\mathbb{R}^d \rightarrow \mathbb{R}$ with a d-dimensional predictor variable $\boldsymbol{X}$ and a 1-dimensional response/target $\boldsymbol{Y}$. A base procedure is a specific algorithm which yields one estimated function $\hat{g}(\cdot)$. It is possible to run different base procedures many times to have different estimated functions. An ensemble-based function estimates $\hat{g}_{en}(\cdot)$ by having a linear combination of the individual function estimates $\hat{g}_k(\cdot)$:

$$\hat{g}_{en}(\cdot)=\sum_{k=1}^{M} c_k \hat{g}_k(\cdot)$$

where c_k are the linear combination coefficients. The coefficients can simply assume averaging weights, or can assume different numerical values.

Here, the ensemble method is outlined for the improvement of the temporal prediction of the ENNs and the SVR. The training stage can be further divided into two steps. Firstly, a number of individual ENN and SVR models of different structure parameters are tested against their predictability of the microarray time series. Secondly, the ensemble of these ENN and SVR models is constructed. In this ensemble step, the models with poor predictability are discarded, while the best predictors of the individual ENN and SVR models are selected for the ensemble. This can reduce the effects of the models with poor training performance. To simplify the ensemble model, all of these selected models have equal weight. Then, in the testing stage, these selected models are fed with the testing dataset and produced their predicted values individually. The median of these predicted values is adopted as the prediction of the ensemble method. A median is a statistical number separating the higher half of a sample from the lower half. From our experience, the median is less affected by the extreme values of a dataset than the case with another statistical number – the mean.

5.3.2.4 Pedagogical Rule Extraction for Biological Network Inference

A rule extraction technique can be used to infer the output of the ensemble method. Generally speaking, there exists a causality relation between two genes if the expression level of one gene is predictive of the expression level of the other gene concerned (Mukhopadhyay and Chatterjee 2007). The causality problem is to study the interaction among genes and establish a directional pattern for the actions of activating and inhibiting among the genes in the network. Traditional interaction studies like association, correlation or clustering are popular tools for the microarray gene expression datasets, but they do not establish this directional pattern. Thus, a two-step test has been proposed for the network inference. The first step is the causality test, which checks the predictability of the input genes for the target genes. Then, the Pearson correlation test is to establish the directional pattern between the causal gene inputs and their predicted outputs. The experimental results show that the regulatory network can be identified correctly.

When the regulatory relationship is positive, it is called activation. A negative regulatory relationship is called inhibition. A gene regulatory network can then be represented by a directed graph, with nodes representing the genes and links representing the regulatory relationships. Two nodes are said to be without direct regulatory relationship (causality) if there is no direct link between them. Two nodes are said to have an indirect regulatory relationship if there exists an indirect link between the two nodes. The inference of this gene regulatory network can help systems biology researchers to gain deeper knowledge of the gene interaction mechanism for a specific part of the genome. As said in the introduction, various reverse engineering methods for gene regulatory networks have been proposed, for example, Boolean methods (Shmulevich et al. 2002; Lahdesmaki et al. 2003), Bayesian methods (Friedman et al. 2000; Zou and Conzen 2005), regulation matrix methods (Chen et al. 1999), and machine learning approaches like genetic algorithms (Iba and Mimura 2002) and genetic programming (Sakamoto and Iba 2001).

The low explanation capability and lack of transparency of the recurrent neural networks have made neural networks not very suitable for the inference problems, even though the networks have desirable properties like dynamic mapping, universal approximation and computation power for their application to microarray data sets. To address this weakness, various rule extraction techniques have been proposed to augment neural networks with explanation facilities that can be used for a general inference problem. These rule extraction techniques can be categorized according to the granularity of the underlying neural network. Andrews et al. (1995) proposed three identifiers i.e., decompositional, eclectic, and pedagogical, for defining reference points on the spectrum of such perceived degrees of granularity. Pedagogical techniques extract the global relationships between the inputs and the outputs of the neural networks directly, without considering the detailed characteristics and architectures of the underlying neural networks. The outlined network inference technique in this chapter can be regarded as a member of this pedagogical group.

In this outline, a two-step rule extraction procedure has been presented. In the first step, the causality test has been applied to check the input genes' predictability for the target gene expression time series. The causality means that an input variable is helpful for the numerical prediction of a target variable. Generally speaking, a gene G1 is said to be a cause for gene G2 if expression of G1 is predictive of expression of G2 (Mukhopadhyay and Chatterjee 2007). A direct causal relationship can be represented as $G1 \rightarrow G2$, meaning that expression of G1 predicts expression of G2. An indirect causal relationship is a link from G1 to G2 through a sequence of direct causal relationships with at least one intermediate genes G3,…,Gk, and can be represented as $G1 \rightarrow G3... \rightarrow Gk \rightarrow G2$. In our study, specifically, an input gene is said to have the most significant causal relationship with a target gene if that input gene expression can provide the most accurate prediction for the target gene expression. From our experience, the experiment results in the later sections show that this selection procedure can filter out the direct causal relationships against the indirect casual relationship. In our approach, firstly the neural network models are fed with the expression values of genes in turn, and, then the ensemble of these models is formed as followed: Different ENN models have been set up and gone through the above process and produced their recommended activators and inhibitors for the gene regulatory networks. To combine these recommendations together, for each target gene in turn, the mode of the recommended activator/inhibitor gene candidates given by the ENN models is used as the recommendation of the ensemble method. The mode is a statistical value that takes the most frequent value assumed in a sampling of a random variable. The output of the ensemble is checked against the expression values of target gene. From the above definition of the causality, it can be seen that the gene expressions of the best predictability for the target gene have the causality relationship with the target genes.

With the above causality relationship, the directional pattern is still not known yet. Thus, in the second step, the Pearson correlation test has been employed for checking the directional pattern between the causal input genes and the predicted outputs from the first step. The correlation test can show the directional pattern of the influences of the input genes on the target gene as well. The Pearson correlation coefficient is defined as $r_{xy} = \sum(x_i - \bar{x})(y_i - \bar{y})/(n-1)s_x s_y$, where x_i and y_i are the n measurements of two random variables, with $i = 1, 2, \ldots, n$. Here, $\bar{x}$ and $\bar{y}$ are the sample means of these two variables, and s_x and s_y are their sample standard deviations. After computing the Pearson correlation coefficient, a positive correlation coefficient means activation effects, while a negative correlation coefficient means inhibition effects. Our proposed two-step technique has the flexibility of pedagogical group and can easily be adopted for our proposed ensemble of ENNs and SVRs.

5.4 Models for Biological Network Inference

5.4.1 Biological Time Series Datasets

Various biological time series datasets are produced from experimental works or observations of the biological systems. The time series can vary significantly in length and size. Datasets of a ecosystem may span multiple years and serve as powerful tools for scientists to understand the ecosystem function, variation and resilience. Time series gene expression experiments can let the scientists to study a wide range of biological systems under a variety of conditions. The regulatory relationships can be inferred with time lag analysis. Shi et al. (2007) proposed an associated algorithm to address the problem of inferring time-lagged regulatory relationships from multiple time series experiments with varying time-scales. A set of known interacting pairs was used to compute a temporal transformation between every two datasets. A much lower false-positive rate can be obtained.

5.4.2 Analysis with Simulated Non-stationary Datasets

In the studies of the biological time series, it may be sometimes difficult and expensive to obtain the experimental datasets under various conditions. The simulations of the non-stationary datasets can provide a convenient and economical way for the researchers, provided that the simulations are consistent with the knowledge about the real biological events. In many studies, the scientists use both the simulated and the experimental datasets such that the results can cover not only a wide range of simulated cases but also a few real cases. In the study of non-stationarity of circadian rhythms, Refinetti (2003) used the computer to generate the simulated datasets to evaluate the extent of non-stationarity of circadian rhythms. Experimental data sets of the rhythms of body temperature and running-wheel activity in several rodent species were also studied. The results show that the robustness of a circadian rhythm can serve as an index of the stationarity of the rhythm. In the spectral analysis of short-term recordings of cardiovascular variability signals (CVSs) (Pinna et al. 1996), synthetic CVSs were generated as AR processes whose parameters were estimated on corresponding time series of normal, post-myocardial infarction and congestive heart failure subjects. The simulated CVSs time series datasets were then used for the evaluations of the common models and for the model order selection in AR estimation.

In the study of the nonlinear similarity measure for multichannel biological signals (Xu et al. 2007), both simulated data and multichannel electroencephalogram (EEG) signals during behavior studies were used to evaluate the performance of the new measure versus the well established correlation coefficient. In the study of the causality network among a set of gene expression time series, Mukhopadhyay and Chatterjee (2007) conducted the simulation of a number of pathways of the gene expression time series for their proposed causality and pathway search algorithm. In their simulation, a network of 14 genes consisting of one complex and one simple causality relationship network was simulated. The independent genes in the network were all AR(1) process with autocorrelation < 1, while the Granger caused time series are generated with one or more of the causing series, all with a lag of 1 and autocorrelation <1. The causality was established by Granger causality, and the constructing the pathway was achieved by finding the Minimal Spanning Tree within each connected component of the inferred network. The simulation was replicated 100 times to evaluate the general behavior of the constructed graph and the minimal spanning tree.

5.4.3 Analysis with Real Biological Datasets

Even though the costs of preparing the simulated time series are usually low, the real biological datasets are still needed in order to validate the proposed models with real biological cases. Gene expression time series datasets can be obtained from microarray experiments and used for a wide range of biological systems.

It is reported that more than 80% of all time series expression datasets are short with 8 time points or fewer (Ernst et al. 2005). The large number of genes profiles and the small number of time points present unique challenges for the time series analysis, as many patterns are expected to arise at random. Ernst et al. proposed an algorithm specifically designed for clustering short time series expression data. Genes were assigned to a predefined set of model profiles that capture the potential distinct patterns. Both simulated and real biological data were used to show that the algorithm can correctly detect the temporal profile of relevant functional categories. Both the datasets and the Java implementation are available at: http://www.cs.cmu.edu/~jernst/st/.

In the commentary by Wang et al. (2008), the recent advances in methods for short time series analysis including simplification-based approaches and the integration of multi-source information are discussed. Simplification algorithms work by reducing the time series data from continuous to discrete representations prior to analysis. The raw temporal profiles are transformed into a set of symbols or nominal values that can categorize qualitatively the gene expression data into different states or trends. In the works of Di Camillo et al. (2005), the gene expression value is quantized into three patterns of states, i.e., under-expressed, not differential expressed or over-expressed with respect to a baseline pre-defined by a hypothetical distribution. With the discretization, the Dynamic Bayesian Network algorithm was shown to perform better in terms of precision in reconstructing the regulatory network. Incorporating multi-source information is another approach to address limited sampling and to improve the computation analysis of short time series microarray data. Multi-source information includes, for example, prior knowledge, different levels of information, and additional time series datasets from other sources. A major challenge for this approach is how to integrate these different heterogeneous datasets.

5.4.4 Rule Extraction for Reverse Engineering of Biological Networks

Constructing biological networks is one of the important tasks in systems biology. It can help the understanding of biological process and the organizational principles of the biological systems. Regulatory gene networks can be constructed with the gene expression time series. The reverse engineering of gene networks is an important step in the study of complex biological systems. Krishna et al. (2005) applied the sensitive analysis to trained perceptions to reverse engineer the gene networks. A trained neural network is shown to be able to describe a gene network with minimal re-interpretation, with the sensitivity between nodes reflecting the probability of one gene affecting another gene in time. The methodology was tested with both simulated temporal data and rat CNS development time series data. Lee and Yang (2008) applied two intelligent computing techniques, genetic programming and neural computation, to infer two kinds of network models that use continuous variables. The intelligent approach can automate the procedure to construct the network from temporal data.

Li et al. (2006a) proposed a combined literature mining and microarray analysis approach to construct gene networks. A global network is firstly constructed with the literature-based co-occurrence method. Then microarray data is used to refine the network with a multivariate selection procedure. The approach was shown to be reliable in dealing with multiple levels of KEGG gene and pathway. Auliac et al. (2008) investigated the behavior of an evolutionary approach by niching the learning of gene regulatory networks with Bayesian Network (BN). Evolutionary algorithms were used to stochastically evolve a set of candidate BN structures and find the model that best fits data without prior knowledge. A niching strategy was used to reinforce diversity through the population and avoid the trapping of the algorithm in a local minimum during the learning process.

5.5 Discussions and Further Works

In our chapter, an ensemble of Elman neural networks and support vector machines has been described for the gene regulatory networks. Firstly, in the following two sub-sections, recurrent Elman neural networks and support vector machines will be proposed for continuous temporal gene expression modeling. Secondly, an ensemble of these individual ENNs and SVRs models has been proposed for further improving the numerical performance of the individual ENN and SVR models. Lastly, to address the transparency weakness of the neural network, a pedagogical rule extraction has been introduced for the output of the ensemble for regulatory network inference.

One of the focuses of the previous studies is on the oscillations of the cell cycle. Oliva et al. (2005) have investigated the slight oscillations in expression as a function of the cell cycle with the network component analysis (NCA). Another focus is on the GRN problems. The regulation matrix approach has employed the linear and pseudo-linear mathematical models like ordinary and partial differential equations for the network reconstruction. As most biomolecular networks have nonlinear dynamics properties, there exist difficulties in the estimation of parameters for these models. Boolean methods apply the Boolean logic to the discretised gene states for inferring the GRNs. During this discretisation process, some of the information about the mRNA expression levels may be lost and may result in unrealistic inference outputs. Bayesian methods utilize the Bayes' rule to infer the causal relationship between genes. Li et al. (2006) applied the decision tree as the learner in the identification of biological regulatory rules. These approaches do not provide numerical prediction for continuous microarray time series. While the machine learning approach is less restrictive, these methods often lack the transparency for the inference outputs and the robustness for the noise levels of the gene expression experiments.

In summary, the ensemble approach has been outlined for the reverse engineering of GRNs with continuous microarray time series expression levels, to better improve the transparency and robustness of the recurrent neural networks. The potential contributions of our proposed method are twofolds. Firstly, the numerical expression levels of the target genes have been estimated with their regulatory genes. This can be helpful for the testing of new drugs, for example, estimating the numerical effects of changes of regulatory genes on the target genes. The prediction results on the real biological data sets and the simulated non-stationary data sets support the performance of the numerical prediction of gene expression time series with our proposed method. The proposed approach outperforms previous results obtained using other computational techniques for prediction on microarray data sets. Secondly, the proposed rule extraction technique can provide the network inference of the ensemble. The reconstructed GRNs can enable us to better understand the dynamics of the underlying biological processes. The network inference results demonstrate that the gene regulatory network can be reconstructed satisfactorily.

A major problem of the reconstruction of the GRNs is the small number of sampling while at the same time the large number of genes. This can be addressed by applying a preprocessing step to cluster the gene expression data. Transcript factor activities (TFAs) from various types of DNA microarray data are demonstrated to be useful in inferring transcription factor functions and interactions in Saccharomyces cerevisiae cell cycle regulation (Yang et al. 2005). The transcript factor information, binding motifs and the time series expression data can be combined in the clustering process to find the transcriptional modules, for example in the Kundaje study (Kundaje et al. 2005). The issue can also be addressed with statistical and computational intelligent methods. For example, the results of the neural-genetic hybrid proposed by Keedwell and Narayanan (2005) showed that those genes significantly involved in interaction with other genes can be highlighted and hypothetical gene networks proposed for further testing. Chan et al. (2008) showed that the least angle regression (LARS) and autoregressive (AR) methods can successfully extracted biologically relevant regulatory networks for further analysis. In our present study, we have limited the number of genes considered to highlight the validation of our methodology with the results by other previous works. Through the above gene selection methods, subsets of genes under specific regulatory processes may be identified and can be further investigated with our proposed ensemble method, while the ensemble approach has been shown to be able to produce the more accurate forecast in studies like this one and other forecasting applications as well (Maqsood and Abraham 2007).

Chapter 6
Real-Word Application III: Developing Innovative Computing Algorithms for Astronomical Time Series

Abstract With the advances of the technologies for the sky surveys, massive amount of survey data become available. It would be very helpful for the automatic and semi-automatic methods in the classifications/detections of the astrophysical objects. In fact, for surveys of millions of objects, it may not be possible to detect the desired objects by expert inspection alone. Quasars are interesting astrophysical objects that have been recently discovered more comprehensively from the sky surveys. Automatic and semi-automatic methods have been proposed for the detection of the quasars from the massive data produced by the modern sky surveys. In this chapter, the first section describes about the existing automatic and semi-automatic methods for the comprehensive search of quasars. Secondly, some innovative computing algorithms are described about how to classify the light curves of the quasars against light curves of the other stars.

Keywords Innovative computing • Real-world applications • Astronomical time series • Quasar • Classifications • Detections

6.1 Astronomical Time Series

The astronomical time series include temporal data of astronomical objects like interacting binaries, planet searches, pulsars, active galactic nuclei, quasars, and gravitational lensing, etc. The advances of the technologies for the sky surveys significantly increase the quantity, quality and variety of the time series data of the astronomical events (Maoz et al. 1997). As a result, higher and higher levels of sophisticated methods are required for the astronomers to analysis these more and more complex temporal data.

S.-I. Ao, *Applied Time Series Analysis and Innovative Computing*,
Lecture Notes in Electrical Engineering 59, DOI 10.1007/978-90-481-8768-3_6,

6.2 Advances and Applications of Innovative Computing Paradigms

6.2.1 Classification of Astronomical Time Series

In most empirical sciences, classification usually plays an important role in the early phases of the studies. Astronomers consistently need to classify the observed astronomical time series. Traditionally, the classification is conducted by the human expert's unaided eye–brain. Nowadays, a variety of classification methods have been proposed for the astronomical time series. The goal is to create a system that can classify the astronomical time series automatically with the performance as well as human experts. Wolf et al. (2001) proposed a photometric method for the classification of stars, galaxies and quasars in multi-color surveys. The method can extract the information content of object colors in a statistically correct way, and conduct the classification for galaxies and quasars in a unified approach based on the same probability density functions. Automated classification system for the spectral classification has been sought for over 40 years. Weaver (2000) proposed an artificial neural network model to conduct two-dimensional spectral classification of the components of binary stars. The ANN model can classify the spectra based on the 15 Å resolution near-infrared spectral classification system. The proposed ANN model was applied to seven binary stars and the results showed that the mean absolute errors could be reduced to about 0.5 types in temperature and 0.33 classes in luminosity.

6.2.2 Clustering of Astronomical Time Series

A common task of the astronomical time series is to find the patterns in the temporal data. Clustering methods can efficiently find patterns that do not rely on a regular periodicity. Perlman and Java (2003) developed a Java toolbox of clustering method for astronomical time series. The toolbox consists of a rule inference engine, programmed in C and a visualizer, programmed in Java. A sliding window was applied to the time series to get subsequences. Then, a greedy clustering algorithm was applied to cluster these points. Once a good fit was achieved, the clusters were considered as the basic shapes of the time series, with the entire series composed of combinations of these basic shapes. Interesting probabilistic rules could be found between the clusters in two time series. Shin et al. (2009) proposed a nonparametric Bayesian clustering algorithm to detect various types of variable objects within massive astronomical time series. It extracts information from a given dataset, which is described by six variability indices. The Gaussian Mixture Model was applied to recover the clusters that are described by six-dimensional multivariate Gaussian distributions. The approach is effective for surveys like GAIA, Pan-Stars, and LSST, which produced massive time series data.

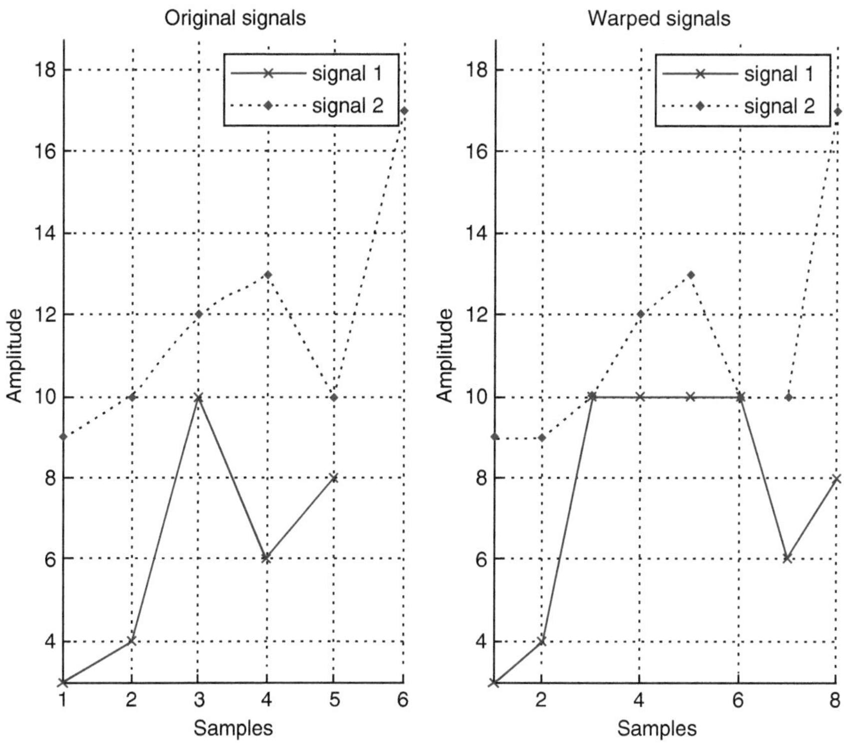

Fig. 6.1 The original signals (*left*) vs. the warped signals by the dynamic time warping method (*right*)

Dynamic time warping (DTW) is a useful tool to measure the similarity between two time sequences. It can allow a non-linear mapping of one sequence to another by minimizing the distance between the two (Ratanamahatana and Keogh 2005). DTW method can be applied for various tasks in time series problems like clustering, classification and anomaly detection etc. An advantage of DTW is its flexibility to allow two sequences that are similar but locally out of phase to align in a non-linear way. As an example, for two sequences of signal 1: [3 4 10 6 8] and signal 2: [9 10 12 13 10 17], the computed optical alignments are [3 4 10 10 10 10 6 8] and [9 9 10 12 13 10 10 17] respectively, as shown in Fig. 6.1.

6.2.3 Semi-Supervised Learning for Astronomical Time Series

Broadly speaking, machine learning falls into the categories of supervised learning and unsupervised learning. An interesting research area is attempts to combine the strengths of both supervised and unsupervised learning methods. In semi-supervised learning, both labeled and unlabeled data are used in the training (Chapelle et al. 2006). The use of a small amount of labeled data can improve considerably the learning accuracy. Bazell et al. (2003) explored the use of unlabeled data for supervised classification and for semi-supervised clustering. Methods were examined whereby supervised classification techniques could use large volumes of unlabeled data to improve the ability for classification. Expectation Maximization (EM) algorithm was used to iteratively refine estimates of model parameters when using incomplete data.

6.2.4 Anomaly Detection of Astronomical Time Series

The works of finding unusual time series are attracting more attention. The current state is that many existing approaches require that the time series can fit in the main memory of the computing system. Nevertheless, multi-terabyte time series datasets are the norm in astronomy. Yankov et al. (2008) proposed a disk aware algorithm to discover the time series discord. The method is exact and only two linear scans of the disk with a tiny buffer of main memory. It is used to provide further evidence of the effectiveness of the discord definition in areas like astronomy. Multi-gigabyte datasets containing tens of millions of time series of star light curve data were used to show that the approach can efficiently mine the unusual time series. Vestrand et al. (2004) studied the autonomous robotic instrumentation with the ability to extract pertinent information from the data stream in real time. The integration of the robotic telescopes with modern artificial intelligence techniques can lead to discovery engines that are capable of autonomously mining the sky in real time. Such a system may be useful for global census of stellar flaring, finding unpredicted close encounters with nearby solar system objects, searching for "orphan" gamma-ray bursts, and discovering novae etc.

6.3 Motivations for Investigating the Quasar Time Series with Innovative Approaches

Quasars, quasi-stellar radio sources, are powerfully energetic and distant galaxy with an active galactic nucleus. Most quasars are farther than 3 billion light-years away from the Earth, but they are so luminous enough to be still visible. Their luminosity was found to vary on a variety of time scales. As the quasars are all very old, the studies of quasars may help us better understand the early stage of the universe. The quasars can also serve as suitable reference frames against which to measure the proper motion of other astrophysical objects (Geha et al. 2003). They may also be helpful in the absorption line studies of the interstellar medium in the sky region. It is interesting itself to investigate the physical mechanisms underlying quasars light variation. Thus, innovative approaches have been applied for the investigation of the quasars light variation, which is shown to be helpful for the automatic quasar selection (Ao 2009).

6.4 Advances in Emerging Methods for Quasar Studies

6.4.1 *Variability Properties of the Quasar Light Curves*

In the work of Scholz et al. (1997) on the combined variability-proper motion survey for QSOs on Schmidt plates, in order to search for objects with variability timescales longer than a few months, the mean structure functions for time-lags longer than 100 d were compared with that for time-lags shorter than 100 d. The ratio $RS_{100} = SF\,(\tau > 100) \,/\, SF\,(\tau < 100)$ serves as an indicator of long-term variability. The criterion $\log_{10}(RS_{100}) > 0.15$ was adopted, which yielded a strong reduction of the candidate sample while only a few QSOs would remain undetected. The final sample contained 168 QSO candidates with estimated success of over 40% and completeness about 90%. Here, the variability index is defined as:

$$I^{(B)}_{\mathrm{var},i} = \frac{1}{n^i_{pl}} \sum_{j=1}^{n^i_{pl}} \frac{\left|B_{ij} - \overline{B}_i\right|}{\sigma_j\left(B_{ij}\right)} \text{ with } \overline{B}_i = \frac{1}{n^i_{pl}} \sum_{j=1}^{n^i_{pl}} B_{ij}$$

where $\sigma_j\,(B_{ij})$ is the photometric random error of the *j*th plate at the magnitude *B* measured for the object on this plate. The proper motion index was defined as

$$I_{pm} = \frac{\mu_x^2 + \mu_y^2}{\sqrt{\mu_x^2\sigma_{\mu_x}^2 + \mu_y^2\sigma_{\mu_y}^2}}$$

where (μ_x, μ_y) are the two proper motion components and $(\sigma_{\mu_x}, \sigma_{\mu_y})$ are their errors.

Meusinger et al. (2002) conducted a combined variability and proper motion (VPM) survey on the globular cluster M 3. In their spectroscopic follow-up observations of QSO candidates, 114 QSOs and 10 Sey1s were pre-estimated with 90 completeness limit of the survey. In the VPM survey, the spectral energy distribution of QSOs is not used as a criterion. It is found that these VPM QSOs do not significantly differ from those samples from colour selection or slitless spectroscopy. As the variability of flux densities and stationarity of positions are two fundamental properties of quasars, the VPM survey is based on indices for star-like image structure, positional stationarity, overall variability, and long-term variability measured on 57 B plates taken with the Tautenburg Schmidt plates between 1964 and 1994.

The proper motion index, I_{pm}, is the measured proper motion in units of the proper motion error. The overall variability index, I_{var}, is the deviation of the individual magnitudes about the mean magnitude, and is normalized by the average magnitude scatter for star-like objects in the same magnitude range. I_{ltvar} is the means of structure function analysis and is computed for all star-like objects with $I_{\text{pm}} < 4$ and $B < 20$. The selection thresholds for the indices were based on the statistics of the previously known 90 QSOs in the field. With the set of constraints: $I_{\text{pm}} < 4$, $I_{\text{var}} > 1.3$, and $I_{\text{ltvar}} > 1.4$, the pre-estimated values for the success rate and the completeness are 90% and 40%, respectively. With the spectroscopic follow-up observations of 198 candidates, it is estimated that the completeness of the VPM QSO sample with $B \leq 19.7$ is 94%.

Berk et al. (2004) investigated how quasar variability in the rest-frame optical/UV regime depended on rest-frame time lag, luminosity, rest wavelength, redshift, the presence of radio and X-ray emission, and the presence of broad absorption line systems. A sample of over 25,000 quasars from the SDSS was used. The imaging photometry was compared with three-band spectrophotometry obtained at later epochs spanning time lags up to about 2 years. The results show that time dependence of variability (the structure function) is fitted well with a single power law of an index $\gamma = 0.246 \pm 0.008$, on timescales from days to years. It is found that there exist an anticorrelation of variability amplitude with rest wavelength, and a strong anticorrelation of variability with quasar luminosity. It is also shown that there exists a significant positive correlation of variability amplitude redshift.

The long-term quasar variability has been studied in works like Vries et al. (2005). In Vries' work, a sample of 35,165 quasars from the SDSS Data Release 2, along with 6,413 additional quasars in the same area of the sky from (2dF) QSO Redshift Survey, was used. The authors used the structure function (SF) for the variability diagnostic. The results on the quasar SF showed that most the long-term variations are intrinsic to the quasar itself. The results showed that there was no upper bound to the preferred variability timescale (smaller than a few decades). Second, the magnitude of the quasar variability was found to be a clear function of wavelength. Third, high-luminosity quasars tended to vary less than low-luminosity quasars. The results showed that the quasar variability was intrinsic to the source and was caused by chromatic outbursts/flares. Currently, the model that best explained this observed behavior was based on accretion DIs. In the work, the SF function is defined as

$$S(\tau) = \left\{ \frac{1}{N(\tau)} \sum_{i<j} \left[m(i) - m(j) \right]^2 \right\}^{1/2}$$

with the summation over all the combinations of measurements for with $\tau = t_j - t_i$. Here, the $n(n-1)/2$ permutations are grouped into bins that contain at least 200 measurements. The SF value for each bin is the rms of the magnitude permutations. The SF similarities are quantified in terms of their offset distributions. The mean SF offset and its 1 σ uncertainty, $\bar{O}$ and $\Delta \bar{O}$, are defined as

$$\bar{O} = \frac{1}{N} \sum_{i}^{N} S_A(i) - S_B(i)$$

$$\Delta\bar{O} = \frac{\sigma}{\sqrt{N}} = \frac{1}{N}\left\{\sum_{i}^{N}\left[S_A(i) - S_B(i) - \bar{O}\right]^2\right\}^{1/2}$$

where the SF curves are labeled A and B, both with $N = 42$ bins here, and

$$N \approx N^{1/2}(N-1)^{1/2}.$$

The time lags between the optical continuum bands for a sample of 42 quasars were studied in the work (Bachev 2008). The interpolation cross-correlation function (*ICCF*) method (Gaskell and Sparke 1986) was applied to the quasar sample (Giveon et al. 1999), which was monitored in two colors (B and R). The maximum of $ICCF(\tau)$ showed the time delay between the bands. It was found that most of the objects showed a delay in the red light curve behind the blue one (a positive lag). The lag was about +4 days on average (+3 for the median). The results were broadly consistent with the reprocessing model.

In Sections 6.4.2 and 6.4.3, several existing automatic and semi-automatic methods for the detection of the quasars from the massive data are reviewed, along with brief summaries of their results and limitations. In Section 6.4.4, some machine learning algorithms are explored for the automatic classification of the light curves of the quasars against the very similar light curves of the other stars. In Section 6.4, experimental results will also be shown for this exploratory work.

6.4.2 Algorithms Based on Variability and Proper Motion for Quasar Classification

Geha et al. (2003) reported that 47 spectroscopically-confirmed quasars were discovered behind the Magellanic Clouds via photometric variability in the MACHO database. Thirty-eight quasars are behind the Large Magellanic Clouds (LMC) and nine behind the Small Magellanic Clouds (SMC). The follow-up spectroscopic detection efficiency was 20%, with emission-line Be stars as the primary contaminants. The first 5.7 years of MACHO data were used for the selection of the quasar candidates. The final 7.5 year MACHO light curves were used in the lightcurve analysis of quasars. Previously 9 Active Galactic Nuclei (AGN) in MACHO database were cross-identified with X-ray sources by Schmidtke et al. (1999), and one AGN by Blanco and Heathcote (1986). Two additional sources in the MACHO database were presented by Dobrzycki et al. (2003) subsequent to candidate selection. The lightcurves for these sources served as a training set for the developing of the selection method. The results showed that a total of 259 candidate quasars were observed for spectroscopic confirmation. Thirty-eight quasars were confirmed behind the LMC and 9 behind the SMC. The quasars cover the redshift interval $0.2<z<2.8$ and mean magnitudes $16.6 \le \bar{V} \le 20.1$. The primary contamination during spectroscopic follow-up were the Be/Ae stars. Of the 258 quasar candidates observed, 188 were the emission-line Be/Ae stars.

The variability selection method for quasar candidate, designed to automatically reject known classes of variable stars, with the final step being a selection by eye, was shown as followed:

6.4.2.1 Data: Deviation from a Constant Brightness Lightcurve

There are 12 million objects in MACHO LMC and 2 million objects in MACHO SMC. From the Level 1 MACHO database, 140,000 objects were flagged as having a significant deviation from a constant brightness lightcurve (Alcock et al. 2000), and the proposed selection method began with these objects. The objects were required to have a minimum of 50 photometric measurements in *V*- and *R*-bands.

Step 1: Magnitude and Color Tests

Weighted average magnitudes were calculated from the standard equation:

$$\overline{V} \equiv \sum_{i=1}^{N} \frac{V_i}{\sigma_{V,i}^2} \Bigg/ \sum_{i=1}^{N} \frac{1}{\sigma_{V,i}^2}$$

where N is the total number of individual photometric measurements V_i with corresponding errors $\sigma_{v,i}$. Candidates were required to be with $16 \le \overline{V} \le 20$ and have weighted average colors bluer than $\overline{(V-R)} \le 1.0$. The color cut eliminated long-period quasi-periodic variable stars, while including all the AGNs in the training set.

Step 2: Statistics Tests of Lightcurve Variability

Two statistics were used to check the amount of lightcurve variability. The intrinsic variability was as

$$\widehat{\sigma_V} \equiv \sqrt{\frac{\sum\left(V_i - \overline{V}\right)^2}{N-1} - \frac{\sum \sigma_{V,i}^2}{N}}$$

It is required that the intrinsic variability of MACHO candidate quasars to be larger than 0.05 magnitudes ($\widehat{\sigma_V} \ge 0.05$), as the sample of 42 quasars by Giveon et al. (1999) were found to be of $0.05 \le \widehat{\sigma_B} \le 0.32$ in the *B*-band. The second variability statistics is the variability index (Welch and Stetson 1993):

$$I \equiv \sqrt{\frac{1}{N(N-1)}} \sum_{i=1}^{N} \left(\frac{V_i - \overline{V}}{\sigma_{V,i}}\right)\left(\frac{R_i - \overline{R}}{\sigma_{R,i}}\right)$$

The index measures the correlated variability between the MACHO *V*- and *R*-bands, and approaches zero for uncorrelated variability. As quasar variability is expected to be highly correlated between the MACHO passbands, the selection criterion is set as $I \ge 1.5$ for quasar candidates.

Step 3: Removal of Periodic Variable Stars

The periodic variable stars, such as Cepheid and RR Lyrae stars, were removed using the MACHO Variable Catalogue. Lightcurves of a period (τ) at high significance in both passbands over the range $0.1 \le \tau \le 500$ days were rejected.

Step 4: Examination by Eye

Candidate lightcurves were examined by eye for removing objects with spurious noise characteristics or quasi-periodic components. About 2,500 light curves were examined by eye, with a total of 360 lightcurves considered candidate quasars.

In addition, the authors have search for optically-variable counterparts in the error boxes of suspected extragalactic radio and X-ray sources in several Magellanic Cloud surveys.

Eyer (2002) proposed variability selection method to find QSO (quasi-stellar object, or quasar) candidates behind the Small and Large Magellanic Clouds (SMC and LMC). The optical variability properties of the point sources from the OGLE-II photometric data were utilized. The LMC catalog has 7 million stars, and SMC data has 2.2 million stars. The variable star catalog listed 53,000 stars for LMC and 15,000 stars for SMC. The overall results were that 118 and 15 QSO candidates are identified for the LMC and SMC, respectively, but, the confirmation with spectroscopic data has not been done yet. The schematic description of the selection criteria was as followed.

6.4.2.2 First Criterion: Selection on Photometry: Magnitude and Colour

The motivation for the selection on photometry is that the color-magnitude for the stars observed by OGLE-II and for the variable stars has different distributions. Objects bluer than $V - I = 0.9$ mag and within 17 mag $< I <$ 20.5 mag were selected. This is to reject the time series of the red giant branch variables stars. From the experience with SDSS, it is estimated that a cut $I >$ 17 mag will exclude one or two QSOs in all fields together. For the LMC data, this magnitude cut selected a population of main sequence blue variables, RR Lyr stars and red giant stars. Some short period Cepheids are also included in the SMC data. The results showed that 6,241 objects are selected for LMC and 1,553 objects for SMC.

6.4.2.3 Second Criterion: The Slope of Variograms

It is known that the QSOs usually have little variability at short time scales. The variability is irregular and aperiodic and increases as longer time-scales are observed.

The variograms/structure functions are employed for the selection of variability for time scales longer than 100 days. The variograms functions are computed as followed: All pair differences of time, $h_{ij} = JD_j - JD_i$, and squared pair differences of magnitude, $(I_j - I_i)^2$, are calculated, where h is the lag and JD is the Julian Date. The median of the subsample, $2\Upsilon_{med}(h)$, of $(I_j - I_i)^2$ formed by all possible pairs i, j, where $h_{ij} < h$. The variogram can estimate the spread of the distribution formed by the $I_j - I_i$. Its slope can serve as a discriminating criterion to select the time series that has increasing variability for longer time scales.

In Eyer's work, the objects with a compute slope of variogram less than 0.1 for $h \geq 100$ days are rejected. The authors established the value for this limit empirically (with the variable QSO, behind LMC, OGLE050833.29-685427.5), and also by conducting Monte Carlo simulations on the SMC data.

6.4.2.4 Third Criterion: QSO and Be Star Colors

With this slope selection criterion, the list of QSO candidates reduced to 649 objects for LMC and 179 for SMC. The selection cut at $B - V = 0.04$ mag can reject 87% of Be stars and reject less than 1% of the QSOs. This selection on color can cut the sample by half to 312 stars.

6.4.2.5 Fourth Criterion: Manual Selection

A few undesirable effects, like spurious monotonic slopes, field overlap, effect of a bleeding column or extended wings, brighter or dimmer step-like variation in the data, etc., were identified manually. Then the authors selected manually the 312 stars on individual basis thereby to eliminate these undesirable effects.

6.4.2.6 Follow-Up Spectroscopy Experiments and Works

Dobrzycki et al. (2003) conducted the follow-up spectroscopic observations of quasars candidates in the SMC selected by Eyer (2002) from the OGLE database. Of the 12 QSO candidates, five are confirmed quasars. Two of these quasars were also independently identified by Geha et al. (2003). The emission redshifts of the five quasars range from 0.27 to 2.17. It shows that Eyer's method is efficient and the contaminants in the list of candidates were predominantly early type stars in the SMC. No obvious trend or difference was observed between the light curves of these quasars and stars. The optical spectra were obtained with the Magellan Baade 6–5 meter telescope. The spectra of all five quasars show at least two emission lines. This can enable the unambiguous determination of emission redshifts.

Dobrzycki et al. (2005) presented their discovery of nine quasars behind the Large Magellanic Cloud. Six of them were identified as part of the systematic variability-based search for QSOs in the objects from the OGLE-II database. This variability-based identification was performed by Eyer (2002), and the list formed the main part of the candidate pool. Several candidates from the X-ray selected sample (Dobrzycki et al. 2002) were also followed up. The remaining three quasars were identified by followup spectroscopy of optical counterparts to X-ray sources.

For the variability selected quasars, during the authors' observing runs, observations of 108 objects out of the remaining 111 candidates were completed. Six of them were identified as QSOs. For the X-ray selected quasars, a total of 30 objects were observed and three was found to be quasars. Combining with previous found quasars from the OGLE-II LMC and SMC databases, there are fourteen quasars selected. From the colour–colour diagram, the authors found that setting a cutoff at $V–I \approx 0.3$ may increase the efficiency of the variability-based selection method, to 40–50%. It was also found that a decent correlation ($R = 0.64$) is seen for quasars.

Sumi et al. (2005) selected 97 QSO candidates via variability from OGLE-II DIA variable star catalogue in the Galactic Bulge fields. Spectroscopic Follow-up observations were planned to confirm the QSO candidates in four of these fields (BUL_SC1, 2, 32 and 45). Sumi et al. used the VI photometric maps of OGLE-II fields (Udalski et al. 2002), and the light curves of objects in the OGLE-II Galactic Bulge variable star catalogue (Wo´zniak et al. 2002). 30,219 objects which are fainter than $I_0 = 16$ mag are selected, where I_0 is the I-band magnitude.

Following the procedure of Eyer (2002), variogram function is used to find objects with increasing variability on longer time-scales. The slope of the variogram is defined as: $S_{var} = dVar(h)/d\log h$, where differences of times, $h_{i,j} = t_j - t_i$. For a given bin of log$h$, $Var(h)$ is the median of $(I_j - I_i)^2$, where $(I_j - I_i)^2$ is square of difference of magnitudes. The AoV algorithm (Schwarzenberg-Czerny 1989) is employed to search the periodicity in the light curves. The light curves in which strong periodicity is detected with AoV periodogram of larger than 50 were rejected, and 3,201 light curves remain. The remaining light curves were inspected by eye. Firstly, the apparent periodic and semi-periodic variables were rejected. Then stars with contamination like proper motion objects and artifacts around bright Variable stars were rejected. Finally 97 QSO candidates were selected.

6.4.3 Bayesian Classification for Efficient Photometric Selection of Quasars

Sandage and Wyndham (1965) pioneered the selection of quasars from multicolor imaging data and researches on this area continued through the years. The proposed selection algorithm seeks to explore all the regions of color space that quasars are known to occupy. In the work of Richards et al. (2002), the quasar candidates were selected via their nonstellar colors in *ugriz* broadband photometry and by matching unresolved sources to the FIRST radio catalogs. Nearly 95% of previously known quasars were recovered, and the overall completeness is expected to be over 90%. The overall efficiency is better than 65%. The SDSS quasar survey will be approximately 4 times the size of the concurrent 2dF QSO Redshift Survey (Croom et al. 2001). The approach is similar to previous studies, but certain characteristics of the SDSS make the selection of SDSS quasars unique.

Briefly, the proposed algorithm works as follows: (1) Objects with spurious fluxes in the imaging data are rejected. (2) Point-source matches to FIRST radio sources (Becker et al. 1995) are preferentially targeted without reference to their colors.

(3) The sources remaining after the first step are compared to the distribution of normal stars and galaxies in two distinct three-dimensional color spaces. Once quasar candidates were identified, the SDSS fiber-fed spectrographs were used to obtain the spectra of each candidate.

In the color selection process, the first step is to remove from consideration the region inhabited by stars. It is known that the colors of ordinary stars occupy a continuous, almost one-dimensional region in $(u - g)$, $(g - r)$, $(r - i)$, $(i - z)$ color–color–color–color space (Newberg and Yanny 1997). As a result of their distinct colors, quasars candidates can be identified as outliers from the stellar locus. The proposed algorithm chooses quasar candidate objects that are more than 4σ from the stellar locus, where σ can be determined from the errors of the objects in question and the width of the stellar locus at the nearest point. The selection is limited to objects fainter than $i^* = 15$. The *ugri*-selected objects are targeted to $i^* = 19.1$, while *griz*-selected objects are targeted to $i^* = 20.2$. Outliers from the stellar locus are targeted for follow-up spectroscopy.

The completeness of the proposed algorithm is estimated by checking with the recovering of the previously known quasars and the selection of the simulated quasars. A region of about 100 deg^2 of sky that has been targeted by the final version of the algorithm has been used to test the efficiency of the algorithm. It is found that 1,113 of the 1,687 quasar candidates with spectra are quasars or AGNs, with an efficiency of 66.0%.

6.4.3.1 Follow-Up Works to This Automatic Photometric Selection of Quasars

In the follow-up work of Richards et al. (2004), a non-parametric Bayesian classification, based on kernel density estimation, was applied to parameterize the color distribution of astronomical sources. A catalog of 100,563 unresolved, UV-excess (UVX) quasar candidates to $g = 21$ from 2,099 deg^2 of the Sloan Digital Sky Survey (SDSS) Data Release One (DR1) imaging data was presented. From the existing spectral of 22,737 sources, 97.6% were found to be quasars. It was estimated that 95.0% of the objects in the catalog are quasars. The Bayes' Rule was applied to weight each likelihood with its corresponding prior probability to compute the posterior probability of an object x being a star or quasar:

$$P(C_1 \mid x) = \frac{p(x \mid C_1)P(C_1)}{p(x \mid C_1)P(C_1) + p(x \mid C_2)P(C_2)}$$

where C_1 and C_2 are the star classes. Here, objects with $P(C_1 \mid x) > 0.5$ are classified as stars, and objects with $P(C_1 \mid x) < 0.5$ are classified as quasars.

6.4.3.2 Nonparametric Bayesian Classification (NBC)

The quasar training set is the four primary SDSS colors (u–g, g–r, r–i, i–z) of the 16,713 quasars from Schneider et al. (2003). The colors from a random sample of 10% of all point sources in DR1 imaging area with $14.5 < g < 21.0$ were also used. Those objects that do not pass the photometric quality tests (Richards et al. 2002) were rejected. The spectroscopically confirmed quasars were rejected too. The total number of objects in the initial training set was 478,144. The test set composed of SDSS-DR1 point sources with $u - g < 1.0$. Those objects that do not pass the photometric quality tests (Richards et al. 2002) were rejected. The full set contained 831,600 objects.

The likelihood of each object x in the test set with respect to each training set was computed with the nonparametric kernel density estimator:

$$\hat{p}(x) = \frac{1}{N}\sum_{i}^{N} K_h\left(\|x - x_i\|\right)$$

where N is the number of data points, $K_h(z)$, the kernel function, satisfied $\int_{-\infty}^{\infty} K_h(z)dz = 1$, h is a scaling factor called the bandwidth, and z is the "distance" between a pointing the test set to a point in the training set. Here, these distances are the 4-D Euclidean color differences, $\|x - x_i\|$, and a Gaussian kernel,

$$K_h\left(\|x - x_i\|\right) = \frac{1}{h\sqrt{2\pi}}\exp - \frac{\|x - x_i\|^2}{2h^2}$$

was used.

6.4.3.3 Fast Algorithms for Computing the Kernel Density Estimate

A naïve algorithm for computing the kernel density estimate at N points among N points would requires N^2 distance operations. Richards et al. used a fast computational algorithm based on space-partitioning trees and principles similar to those used in N-body solvers (Gray and Moore 2003). A second fast algorithm was proposed for quickly finding the higher posterior probability with details by Gray. Its advantage is that it need not estimate the density completely for each object to be classified. The algorithm only needs to maintain upper and lower bounds on the density for reach class. The bandwidth pair was determined with a statistical criterion for optimal classification accuracy, i.e. leave-one-out, cross-validated accuracy here. The proposed fast NBC algorithm can compute the leave-one-out accuracy score for each pair of bandwidths for the two training sets fast. The best bandwidth was found to be 0.15 mag for each training set, and the corresponding accuracies were 94.48% for the quasars and 97.91% for the stars.

The proposed NBC algorithm was applied to identify quasars. Further cleaning was performed by rejecting objects with large KDE stellar probabilities. This resulted in the 100,563 quasar candidates in the catalog. To estimate the completeness of the proposed approach, the authors checked the results with the SDSS-DRI quasars in the Schneider et al. (2003) catalog. It was found that 94.7% of the quasars were recovered with the proposed algorithm. It was estimated that the filtering of objects via their photometric flags (e.g., those with "fatal errors") may have an effect of 5% additional incompleteness. Simulation approach was also employed to evaluate the completeness of the proposed algorithm. It estimated that the approach is generally at least 95% complete between $z = 0.2$ and $z = 2.0$.

To estimate the efficiency of the catalog, the catalog was matched with three spectroscopic databases, i.e. SDSS-DR1, 2QZ NGP catalog (Croom et al. 2004), and SDSS-DR2 (Abazajian et al. 2004). In all there were 22,737 matches to spectroscopically confirmed objects. 97.6% were confirmed to be quasars. The authors expected that the overall efficiency of the catalog would be 95.0%, obtaining 95,502 quasars in all, even though it may have difficulties to extrapolate the efficiency for these confirmed objects to the entire sample. Further improvements for the algorithm may be achieved in the future with regard to high-redshift and extended quasars. The selection to fainter limits may also be extended. The inclusion of properties like radio- and X-ray detections and lack of proper motion into the algorithm may also improve the efficiency.

Another follow-up work is the one by Richards et al. (2009). The proposed Bayesian algorithm was applied to identify 1,172,157 quasars candidates from a sample of over 40 million SDSS point sources. The overall efficiency of the catalog is about 80%, while a UVX subsample with over 500,000 objects has an expected efficiency of over 97%. Cross-comparison with spectroscopically confirmed type 1 quasars in the COSMOS field was conducted. It was found that the sample is at least 70% complete. The photometric redshifts were estimated for the full sample with an expected accuracy of ±0.3 in about 80% of the time.

The photometric imaging data was from the SDSS Data Release 6. Point sources (type=6) with *i*-band magnitudes between 14.5 and (de-reddened) 21.3 ($psfmag_i > 14.5$ && $psfmag_i$-$extinction_i < 21.3$) were used. The 77,429 SDSS-DR5 quasars with spectra (Schneider et al. 2007) were used for the quasar training set. The set was further supplement with the AAOmega-UKIDSS-SDSS (AUS) QSO Survey, which added another 304 spectroscopically confirmed quasars. The quasars of the $z > 5.7$ by the SDSS (Fan et al. 2006) were also included. In addition, the training set included the 920 objects from the cross-comparison of SDSS and *Spitzer* data. The final quasar training set included 75,382 confirmed quasars. The full test set included 44,449,609 objects to be classified. The Bayesian classification algorithm used in the paper is similar to the Paper I (Richards et al. 2004), with a few changes to the procedure. An improvement over the algorithm in Paper I is on the determination of the optimal bandwidth for classification. An initial broad search of possible bandwidths is now attempted, and then a narrower search around the most optimal bandwidth is executed. The final bandwidths were found 0.11 mag for stars and 0.12 for quasars, with the estimated accuracy (completeness) of 92.6% for the quasar training set. Here, the authors classified the objects as low-redshift ($z \leq 2.2$), mid-redshift ($2.2 < z < 3.5$) and high-redshift ($z \geq 3.5$). The motivation is that the distribution of quasar colors changes considerably with redshift.

6.4.4 Machine Learning Paradigms for Quasar Selection

Zhang and Zhao (2008) compared the performance of Bayesian Belief Networks (BBN), Multilayer Perceptron (MLP) networks and Alternating Decision Trees (ADtree) on separating quasars from stars with the database from the 2MASS and FIRST survey catalogs (of near infrared, radio and known catalogs), with accuracy of mostly over 90%. D'Abrusco et al. (2009) proposed a new unsupervised method to perform quasar candidate selection, based on the clustering of data in the parametric space defined by the photometric colours. The method consists of three steps: (i) an unsupervised clustering performed by Probabilistic Principal Surfaces algorithm; (ii) an agglomerative process driven by a measure of negative entropy, and (iii) a fine tuning of the clustering performances through the exploitation of the information contained in the suitably large base of knowledge (BoK) which is used only for labeling purposes. In the first experiment with optical data (the S-A sample), the efficiency is 81.5%, and completeness is 89.3%. In the second experiment with optical and near infrared data (S-UK sample), the efficiency is 92.3% while the completeness is 91.4%. In the third experiment with optical and near infrared data (S-UK sample and BoK), the efficiency is 97.2%, and the completeness is 94.3%. In the fourth experiment with optical colours only (S-UK sample and BoK), the efficiency is 95.4%, and the completeness is 94.7%.

Acknowledgement The author acknowledges that he has talked with his colleague Dr. Pavlos Protopapas, Harvard University, about the properties of the quasars.

Bibliography

Aach, J., Church, G.: Aligning gene expression time series with time warping algorithms. Bioinformatics **17**(6), 495–508 (2001)

Abazajian, K., et al.: Cosmology and the halo occupation distribution from small-scale galaxy clustering in the sloan digital sky survey. Astrophys. J. **625**, 613–620 (2005)

Abbott, E.: Flatland: A Romance of Many Dimensions. Little, Brown, Boston, USA (1899)

Abraham, A., Philip, N.S., Saratchandran, P.: Modeling chaotic behavior of stock indices using intelligent paradigms. Int. J. Neural Parallel Sci. Comput. **11**(1–2), 143–160 (2003)

Aburto, L., Weber, R.: Demand forecast in a supermarket using a hybrid intelligent system. In: Design and Application of Hybrid Intelligent Systems Book Contents, pp. 1076–1083. IOS, The Netherlands (2003)

Acta, A.: Chemometric strategies for normalisation of gene expression data obtained from cDNA microarrays. Anal. Chim. Acta **446**(1–2), 449–464 (2001)

Adamatzky, A.: Information-processing capabilities of chemical reaction-diffusion systems. 1. Belousov-Zhabotinsky media in hydrogel matrices and on solid supports. Adv. Mater. Opt. Electr. **7**(5), 263–272 (1997)

Adleman, L.: Molecular computation of solutions to combinatorial problems. Science **266**(11), 1021–1024 (1994)

Adya, M., Collopy, F.: How effective are neural networks at forecasting and prediction? A review and evaluation. J. Forecasting **17**, 481–495 (1998)

Agrawal, R., Lin, K., Sawhney, H., Shim, K.: Fast similarity search in the presence of noise, scaling, and translation in times-series databases. In: VLDB, September (1995)

Akdemir, B.: Ensemble adaptive network-based fuzzy inference system with weighted arithmetical mean and application to diagnosis of optic nerve disease from visual-evoked potential signals. Artif. Intell. Med. **43**(2), 141–149 (2008)

Alcock, C., et al.: ApJ, **542**, 281 (2000)

Alvarez-Diaz, M., Alvarez, A.: Forecasting exchange rates using genetic algorithms. Appl. Econ. Lett. **10**, 319–322 (2003)

Amaratunga, D., Cabrera, J.: Exploration and Analysis of DNA Microarray and Protein Array Data. Wiley, New York (2004)

Andrews, R., Diederich, J., Tickle, A.: A survey and critique of techniques for extracting rules from trained artificial neural networks. Knowl-Based Syst. **8**(6), 373–389 (1995)

Ao, S.: Analysis of the interaction of Asian Pacific Indices and forecasting opening prices by hybrid VAR and neural network procedures. In: Proceedings of International Conference on Computational Intelligence for Modelling, Control and Automation, Vienna, Austria, February (2003a)

Ao, S.: Automating stock prediction with neural network and evolutionary Computation. Intelligent data engineering and automated learning. In: Proceedings of Fourth International Conference on Intelligent Data Engineering and Automated Learning 2003, pp. 203–210. Springer, Hong Kong, March (2003b)

Ao, S.: Hybrid VAR, neural network, and evolutionary computation for predicting Asian Pacific market lead-lag dynamics. M. Phil. Thesis, The Chinese University of Hong Kong (2003c)

Ao, S.: Using fuzzy rules for prediction in tourist industry with uncertainty. In: Proceedings of the Fourth International Symposium on Uncertainty Modeling and Analysis, pp. 213–218, University of Maryland, College Park, MD, USA, 21–24 September (2003d). IEEE

Ao, S.: Hybrid intelligent system for pricing the indices of dual-listing stock markets. In: Proceedings of IEEE/WIC International Conference on Intelligent Agent Technology, pp. 495–498, Halifax, Canada, 13–17 October (2003e). IEEE

Ao, S.: A framework for neural network to make business forecasting with hybrid VAR and GA components. Eng. Lett. (Int. Assoc. Eng.) **13**(1), 24–29 (2006)

Ao, S.: Data mining algorithms for genomic analysis. Ph.D. Thesis, The University of Hong Kong (2007a)

Ao, S.: Neural network regressions with fuzzy clustering. In: Proceedings of the 2007 International Conference of Information Engineering of World Congress on Engineering 2007, pp. 507–512, London, UK, 2–4 July (2007b). ISBN: 978-988-98671-5-7

Ao, S.: Constructing linkage disequilibrium map with iterative approach. In: Current Themes in Engineering Technologies: World Congress on Engineering and Computer Science. American Institute of Physics (2008a)

Ao, S.: Data Mining and Applications in Genomics. Springer, Netherlands (2008b)

Ao, S.: Hybrid intelligent regressions with neural network and fuzzy clustering. In: Advances in Computational Algorithms and Data Analysis. Springer, Netherlands (2008c)

Ao, S.: Automatic and semi-automatic methods for the detection of quasars in sky surveys. In: Advances in Machine Learning and Data Analysis. Springer, Netherlands (2009)

Ao, S.: A hybrid neural network cybernetic system for quantifying cross-market dynamics and business forecasting. Soft Computing – A Fusion of Foundations, Methodologies and Applications. Springer. 11 March 2010, DOI: 10.1007/s00500-010-0580-4

Ao, S., Ng, M., Ching, W.: Modeling gene expression network with PCA-NN on continuous inputs and outputs basis. In: Current Trends in High Performance Computing and Its Applications. Proceedings of High Performance Computing and Applications 2004, pp. 209–214, Shanghai, China, 8–10 August 2004

Ao, S., Ng, M., Sham, P.: Constrained unidimensional scaling. In: Programme and Abstracts, 3rd World Conference on Computational Statistics and Data Analysis, International Association for Statistical Computing, Cyprus, pp. 49, (2005)

Ao, S., Yip, K., et al.: CLUSTAG: hierarchical clustering and graph methods for selecting tag SNPs. Bioinformatics **21**(8), 1735–1736 (2005b)

Ao, S., Ng, M.: Gene expression time series modeling with principal component and neural network. Soft Comput. – Fusion. Found. Methodol. Appl. **10**(4), 351–359

Ao, S., Ng, M., Sham, P.: Constrained unidimensional scaling with application to genomics. Computational Statistics and data analysis. Off. J. Int. Assoc. Stat. Comput. **52**(1), 201–210 (2007)

Ao, S., Amouzegar, M., Chen, S. (eds.): Current Themes in Engineering Technologies: World Congress on Engineering and Computer Science. American Institute of Physics San Francisco, USA (2008)

Athanas, P., Abbott, A.: Real-time image processing on a custom computing platform. IEEE Comput. **28**(2), 16–24 (1995)

Auliac, C., Frouin, V., Gidrol, X., d'Alche-Buc, F.: Evolutionary approaches for the reverse-engineering of gene regulatory networks: a study on a biologically realistic dataset. BMC Bioinform. **9**, 91 (2008). doi:10.1186/1471-2105-9-91

Azuaje, F.: A computational neural approach to support the discovery of gene function and classes of cancer. IEEE Trans. Biomed. Eng. **48**, 332–339 (2001)

Bachev, R.: Quasar optical variability: searching for interband time delays. Astron. & Astrophys. **493**(3), 907–911 (2009)

Back, B., Laitinen, T., Sere, K.: Neural networks and genetic algorithms for bankruptcy predictions. Expert Systems Appl. **11**(4), 407–413 (1996)

Banko, M., Etzioni, O.: Strategies for lifelong knowledge extraction from the web. KCAP' 07, October 28–31, 2007, Whistler, British Columbia, Canada (2007)

Barrows, A., Powell, D.: Tunnel-in-the-sky cockpit display for complex remote sensing flight trajectories. In: Fourth International Airborne Remote Sensing Conference and Exhibition/21st Canadian Symposium on Remote Sensing, Ottawa, Ontario, Canada, 21–24 June 1999

Bautu, E., Bautu, A., Luchian, H.: An evolutionary approach for modeling time series. In: The 10th International Symposium on Symbolic and Numeric Algorithms for Scientific Computing, IEEE 2008

Baxt, W.G.: Analysts of the clinical variables driving decision in an artificial neural network trained to identify the presence of myocardial infarction. Ann. Emerg. Med. **21**, 1439–1444 (1992)

Bazell, D., Miller, D.J., Borne, K.: Novel approaches to semi-supervised and unsupervised learning. Astron. Data Anal. Softw. Syst. XII ASP Conf. Ser. **295**, 427 (2003)

Becker, R.H., White, R. L., Helfand, D. J.: ApJ **450**, 559 (1995)

Bederson, B., Shneiderman, B.: The Craft of Information Visualization: Readings and Reflections. Morgan Kaufmann, San Francisco (2003)

Beni, G., Wang, J.: Swarm intelligence in cellular robotic systems. In: Proceedings of NATO Advanced Workshop on Robots and Biological Systems, Tuscany, Italy, 26–30 June 1989

Berk, D., et al.: The ensemble photometric variability of 25,000 quasars in the Sloan Digital Sky Survey. Astrophy. J. **601**, 692 (2004)

Blanco, V. M., Heathcote, S.: PASP **98**, 635 (1986)

Boca, A., Park, D.: Myoelectric signal recognition using fuzzy clustering and artificial neural networks in real time. IEEE World Congr. Comput. Intell. **5**, 3098–3103 (1994)

Bortolan, G., Pedrycz, W.: Fuzzy clustering preprocessing in neural classifiers. Kybernetes **27**(8), 900 (1998)

Bose, S.: Contribution of Indian index futures to price formation in the stock market. Money & Finance, pp. 39–56, Feb 2007.

Boser, B.E., Guyon, I.M., Vapnik, V.N.: A training algorithm for optimal margin classifiers. Proceedings of the Annual Conference on Computational Learning Theory, pp. 144–152. ACM, Pittsburgh, PA (1992)

Box, G., Jenkins, G.: Time series analysis: forecasting and control, Revisedth edn. Holden-Day, Oakland, California (1976)

Box, G., Jenkins, G., Reinsel, G.: Time Series Analysis: Forecasting and Control, 3rd edn. Prentice-Hall, USA (1994)

Bracewell, R.: The Fourier Transform and Its Applications, 3rd edn. McGraw-Hill, Boston, 2000

Brockwell, P., Davis, R.: Time Series: Theory and Methods, 2nd edn. Springer, Germany (1991)

Brown, M., et al.: Knowledge-based analysis of microarray gene expression data by using support vector machines. PNAS **97**(1), 262–267 (2000)

Butte, A., et al.: Comparing the similarity of time-series gene expression using signal processing metrics. J. Biomed. Inform. **34**, 396–405 (2001)

Cai, C., Han, L., Ji, Z., Chen, Y.: Enzyme family classification by support vector machines. Proteins **55**, 66–76 (2004)

Chan, Z., Havukkala, I., Jain, V., Hu, Y., Kasabov, N.: Soft computing methods to predict gene regulatory networks: An integrative approach on time-series gene expression data. Appl. Soft Comput. **8**, 1189–1199 (2008)

Chapelle, O., Schölkopf, B., Zien, A.: Semi-Supervised Learning. MIT, Cambridge, MA (2006)

Charbonnier, S., et al.: Trends extraction and analysis for complex system monitoring and decision support. Eng. Appl. Artif. Intell. **18**(1), 21–36 (2005)

Chen, C., Feng, X., Szeto, J.: Identification of critical genes in microarray experiments by a neuro-fuzzy approach. Comput. Biol. Chem. **30**(5), 372–381 (2006)

Chen, K.Y., Wang, C.H.: Support vector regression with genetic algorithms in forecasting tourism demand. Tourism Manage. **28**(1), 215–226 (2007)

Chen, T., Filkov, V., Skiena, S.: Identifying gene regulatory networks from experimental data. Parallel Comput. **27**, 141–162 (2001)

Chen, T., He, H., Church, G.: Modeling gene expression with differential equations. Pacific Symposium on Biocomputing, pp. 29–40 (1999)

Cho, K., Choo, S., Jung, S., Kim, J., Choi, H., Kim, J.: Reverse engineering of gene regulatory networks. IET Syst. Biol. **1**(3), 149–163 (2007)

Cohen, W.: Fast effective rule induction. In: Proceedings of the 12th International Conference Machine Learning (ML-95), pp. 115–123 (1995)

Cooley, J., Tukey, J.: An algorithm for the machine calculation of complex Fourier series. Math. Comput. **19**, 297–301 (1965)

Corbett, J., et al.: Use of a random coefficient regression (RCR) model to estimate growth parameters. BMC Genet. **4**(Suppl 1), S5 (2003). doi: 10.1186/1471-2156-4-S1-S5

Costa, I.G., et al.: A symbolic approach to gene expression time series analysis. Neural Networks 2002 Brazilian Symposium, pp. 25–30 (2002)

Craig, P., Kennedy, J., Cumming, A.: Towards visualising temporal features in large scale microarray time-series data. Information Visualisation, 2002. Proceedings. Sixth International Conference on, 10–12 July 2002

Cristina, A., et al.: Representation of uncertainties in spatial modelling of decision processes in integrated water resources management. In: Improving Integrated Surface and Groundwater Resources Management in a Vulnerable and Changing World (Proceedings of JS.3 at the Joint IAHS & IAH Convention, Hyderabad, India, September). IAHS Publ. **330**, 289–294 (2009)

Croom, S.M., Smith, R.J., Boyle, B.J., Shanks, T., Loaring, N.S., Miller, L., Lewis, I.J.: MNRAS **322**, L29 (2001)

Croom, S.M., Schade, D., Boyle, B.J., Shanks, T., Miller, L., Smith, R.J.: ApJ **606**, 126 (2004)

Davies, B.: Integral Transforms and Their Applications, 3rd edn. Springer, New York (2002)

Dekker, A., Sen, K., Young, M.R.: Equity market linkages in the Asia Pacific region – a comparison of the orthogonalised and generalized VAR approaches. Global Financ. J. **12**, 1–33 (2001)

DeRisi, J., et al.: Use of a cDNA microarray to analyse gene expression patterns in human cancer. Nat. Genet. **14**, 457–460 (1996)

Devlin, B.: Data Warehouse – From Architecture to Implementation. Addison-Wesley, Reading, MA (1997)

Dewey, T.: From microarrays to networks: mining expression time series. Inform. Biotech. Suppl., Drug Discov. Today **7**(20), 170–175 (2002)

Dietrich, C., et al.: Classification of bioacoustic time series based on the combination of global and local decisions. Pattern Recogn. **37**(12), 2293–2305 (2004)

Di Camillo, B., Sanchez-Cabo, F., Toffolo, G., Nair, S.K., Trajanoski, Z., Cobelli, C.: A quantization method based on threshold optimization for microarray short time series. BMC Bioinform. **6**(Suppl 4), s11 (2005)

Dobrzycki, A., Eyer, L., Stanek, K., Macri, L.: Discovery of nine quasars behind the Large Magellanic Cloud. A&A **442**, 495 (2005)

Dobrzycki, A., Macri, L., Stanek, K., Groot, P.: Variability-selected quasars behind the Small Magellanic Cloud. Astron. J. **125**, 1330 (2003)

Dorffner, G.: Neural networks for time series processing. Neural Netw. World **6**(4), 447–468 (1996)

Dorigo, M., Stützle, T.: Ant Colony Optimization. MIT, USA (2004)

D'Abrusco, R., Longo, G., Walton, N.: Quasar candidates selection in the Virtual Observatory era. Mon. Not. R. Astron. Soc. **396**(1), 223–262 (2009)

Elman, J.L.: Finding structure in time. Cogn. Sci. **14**, 179–211 (1990)

Enders, W.: Applied Econometric Time Series. Wiley, USA (1995)

Engelberg, S.: Digital Signal Processing: An Experimental Approach, Chap. 7, pp. 56. Springer, Berlin (2008)

Ernst, J., Nau, G., Bar-Joseph, Z.: Clustering short time series gene expression data. Bioinformatics **21**(Suppl 1), 159–168 (2005)

Etzioni, O., et al.: Unsupervised named-entity extraction from the web: an experimental study. Artif. Intell. **165**(1), 91–134 (2005)

Evans, G., Karwowski, W., Wilhelm, M.: An Introduction to Fuzzy Set Methodologies for Industrial and Systems Engineering. In: Evans, G.W., Karwowski, W., Wilhelm, M.R. (eds.)

Applications of Fuzzy Set Methodologies in Industrial Engineering, pp. 3–11. Elsevier, New York (1986)
Eyer, L.: Search for QSO Candidates in OGLE-II Data. Acta Astronomica **52**, 241 (2002)
Fall, K.: A delay-tolerant network architecture for challenged internets. SIGCOMM'03, Karlsruhe, Germany, 25–29 August 2003
Faloutsos, C., Ranganathan, M., Manolopoulos, Y.: Fast subsequence matching in time-series databases. In: SIGMOD – Proceedings of Annual Conference, May, Minneapolis, USA, 1994
Fan, X., Carilli, C.L., Keating, B.: ARA&A **44**, 415 (2006)
Fayyad, U., et al.: Advances in Knowledge Discovery and Data Mining. AAAI/MIT, Menlo Park, CA (1996)
Fedorowics, J.: Document-based decision support. In: Sprague, R.H. Jr., Watson, H.J. (eds.) Decision Support for Management. Prentice-Hall, New Jersey (1996)
Feuerverger, A., Vardi, Y.: Positron emission tomography and random coefficients regression. Ann. Inst. Statist. Math. **52**(1), 123–138 (2000)
Finch, H.: Comparison of distance measures in cluster analysis with dichotomous data. J. Data Sci. **3**, 85–100 (2005)
Franses, P.: Time Series Models for Business and Economic Forecasting (Themes in Modern Econometrics). Cambridge University Press, UK (1999)
Friedman, N., Linial, M., Nachman, I., Pe'er, D.: Using Bayesian networks to analyze expression data. J. Comput. Biol. **7**(3/4), 601–620 (2000)
Fry, R.L.: The engineering of cybernetic systems. In: Bayesian Inference and Maximum Entropy Methods in Science and Engineering, AIP Conference Proceedings, vol. 617, pp. 497–528 (2002)
Fujimaki, R., Hirose, S., Nakata, T.: Theoretical analysis of subsequence time-series clustering from a frequency-analysis viewpoint. IN: Proceedings of the 2008 SIAM International Conference on Data Mining, Atlanta, Georgia, 24–26 April 2008
Futschik, M., Kasabov, N.: Fuzzy clustering of gene expression data. Fuzzy Systems, 2002. FUZZ-IEEE'02. In: Proceedings of the 2002 IEEE International Conference on, vol 1, pp. 414–419, 12–17 May 2002
Garcia, V., Debreuve, E., Barlaud, M.: Fast k nearest neighbor search using GPU. In: Proceedings of the CVPR Workshop on Computer Vision on GPU, June, Anchorage, Alaska (2008)
Garetti, M., Taisch, M.: Neural networks in production planning and control. Prod. Plan. Control **10**(4), 324–339 (1999)
Gaskell, C., Sparke, L.: Apj **305**, 175 (1986)
Geha, M., et al.: Variability-selected Quasars in MACHO Project Magellanic Cloud Fields. Astron. J. **125**, 1–12 (2003)
Geschwind, D., Gregg, J.: Microarrays for the Neurosciences. MIT (2002)
Ghazavi, S., Liao, T.: Medical data mining by fuzzy modeling with selected features. Artif. Intell. Med. (2008). doi:10.1016/j.artmed.2008.04.004
Giles, J.: Time series analysis testing for two-step Granger noncausality in trivariate VAR models. In: Handbook of Applied Econometrics and Statistical Inference. Marcel Dekker, New York (2002)
Giveon, U., Maoz, D., Kaspi, S., Netzer, H., Smith, P.S.: MNRAS **306**, 637 (1999)
Gnfhth, J.L., D'Agostino, R.R., Selker, H.P.: Statistical regression techniques for the construction, interpretation and resting of computer neural networks. Med. Decis. Making **12**, 343 (1992)
Goh, B.H.: Evaluating the performance of combining neural networks and genetic algorithms to forecast construction demand: the case of the Singapore residential sector. Constr. Manage. Econ. **18**, 209–217 (2000)
Goldberg, D.E.: Cenetic Algorithms in Search, Optimization and Machine Learning. Addison-Wesley, Reading, MA (1989)
Granger, C.: Investigating causal relations by econometric models and cross-spectral methods. Econometrica **37**, 424–459 (1969)
Gray, A.G., Moore, A.W.: Nonparametric density estimation: toward computational tractability. In: SIAM International conference on Data Mining (2003)
Greene, W.: Econometric Analysis. Prentice-Hall, USA (2000)

Guo, J., Chen, H., Sun, Z., Lin, Y.: A novel method for protein secondary structure prediction using dual-layer SVM and profiles. Proteins **54**, 738–743 (2004)

Guyon, I., Weston, J., Barnhill, S.: Gene selection for cancer classification using support vector machines. Mach. Learn. **46**, 389–422 (2002)

Hamilton, J.: Time series analysis. Princeton University Press, Princeton, NJ, USA (1994)

Hansen, J.V., McDonald, J.B., Nelson, R.D.: Time series prediction with genetic-algorithm designed neural networks: an empirical comparison with modern statistical models. Comput. Intell. **15**, 3 (1999)

Hansen, L., Salamon, P.: Neural network ensembles. IEEE Trans. Pattern Anal. **12**(10), 993–1001 (1990)

Harmon, J., Gross, A.: The Scientific Literature: a Guided Tour. The Chicago University Press, Chicago, USA (2007)

Hartikainen, J., et al. Short-term measurement of heart rate variability. In: Clinical Guide to Cardiac Autonomic Tests. Kluwer, Dordrecht (1998)

Hastings, S., et al. Image processing for the grid. In: Third IEEE International Symposium on Cluster Computing and the Grid (CCGrid'03). Tokyo, Japan, 12–15 May 2003

Hathaway, R., Bezdek, J.: Switching regression models and fuzzy clustering. IEEE Trans. Fuzzy Sys. **1**, 195–204 (1993)

He, L.T.: Time variation paths of international transmission of stock volatility – US vs Hong Kong and South Korea. Glob. Financ. J. **12**, 79–93 (2001)

Herbst, L.: Periodogram analysis and variance fluctuations. J. Roy. Stat. Soc. Series B (Methodological) **25**(2), 442–450 (1963)

Heylighen, F., Joslyn, C.: Cybernetics and second-order cybernetics. In: Encyclopedia of Physical Science and Technology, 3rd edn. Academic, New York (2001)

Hill, T., Lewicki, P.: Statistics Methods and Applications. StatSoft, Tulsa, OK (2007)

Holland, J.H.: Adaptation in Natural and Artificial Systems. University of Michigan Press, Ann Arbor, MI (1975)

Hoppner, F., Klawonn, F., Kruse, R., Runkler, T.: Fuzzy Cluster Analysis: Methods for Classification, Data Analysis and Image Recognition. Wiley, USA (1999)

Horn, D., Axel, I.: Novel clustering algorithm for microarray expression data in a truncated SVD space. Bioinformatics **19**(9), 1110–1115 (2003)

Horn, D., Gottlieb, A.: Algorithm for data clustering in pattern recognition problems based on quantum mechanics. Phys. Rev. Lett. **88**, 018702 (2002)

Huang, J.: Extensions to the k-means algorithm for clustering large data sets with categorical values. Data Min. Knowl. Disc. **2**, 283–304 (1998)

Huang, J., Ng, M.: A fuzzy k-modes algorithm for clustering categorical data. IEEE Trans. Fuzzy Syst. **7**, 446–452 (1999)

Hunter, J., McIntosh, N.: Knowledge-based event detection in complex time series data. In: Joint European Conference on Artificial Intelligence in Medicine and Medical Decision Making, pp. 271–280. Springer, LNCS 1620, 1999

Iba, H., Mimura, A.: Inference of a gene regulatory network by means of interactive evolutionary computing. Inform. Sci. **145**, 225–236 (2002)

Ido, P., Oded, M., Irad, B.: Evaluation of gene-expression clustering via mutual information distance measure. BMC Bioinform. **8**, 111 (2007). doi:10.1186/1471-2105-8-111

Jacobsson, H.: Rule extraction from recurrent neural networks: a taxonomy and review. Neural Comput. **17**, 1223–1263 (2005)

Jajuga, K., Sokolowski, A., Bock, H.: Classification, Clustering, and Data Analysis: Recent Advances and Application. Springer, Germany (2002)

Jeong, B., Jung, H.S., Park, N.K.: A computerized causal forecasting system using genetic algorithms in supply chain management. J. Syst. Softw. **60**, 223–237 (2002)

Ji, X., et al.: Mining gene expression data using a novel approach based on Hidden Markov Models. FEBS Lett. **542**, 124–131 (2003)

Jiang, D., Pei, J., Zhang, A.: DHC: a density-based hierarchical clustering method for time series gene expression data. Bioinformatics and Bioengineering, 2003. Proceedings. Third IEEE Symposium on, 10–12 March 2003, pp. 393–400 (2003)

Jin, T., Ju, J., Sheng, X.: Admire – a prototype of large scale e-collaboration platform. In: Grid and Cooperative Computing. Springer, Germany (2004)

Jing, X., Zhang, D.: A face and palmprint recognition approach based on discriminant DCT feature extraction. IEEE Trans. Syst. Man Cyb. – Part B: Cyb. **34**(6), 2405–2415 (2004)

Jo, H., Han, I., Lee, H.: Bankruptcy prediction using case-based reasoning, neural network and discriminant analysis. Expert Systems Appl. **13**(2), 97–108 (1997)

Joachims, T.: Making large-scale SVM learning practical. In: Schölkopf, B., Burges, C., Smola, A. (eds.) Advances in Kernel Methods – Support Vector Learning, MIT (1999)

Jong, K.: Evolutionary computation: a unified approach. MIT, Cambridge, MA, USA (2006)

Joslin C., et al.: Advanced real-time collaboration over the internet. In: Proceedings of the ACM Symposium on Virtual Reality Software and Technology, pp. 25–32, Seoul, Korea (2000)

Kang, S.: An investigation of the use of feedforward neural networks for forecasting. Ph.D. Thesis, Kent State University (1991)

Keedwell, E., Narayanan, A.: Discovering gene networks with a neural-genetic hybrid. IEEE/ACM Trans. Comput. Biol. Bioinform. **2**(3), 231–242 (2005)

Keogh, E., Pazzani, M.: An enhanced representation of time series which allows fast and accurate classification, clustering and relevance feedback. In: Proceedings of the Fourth International Conference on Knowledge Discovery and Data Mining (KDD-98), New York City, New York, 27–31 August 1998

Keogh, E., et al.: Locally adaptive dimensionality reduction for indexing large time series databases. In: Proceedings of ACM SIGMOD Conference on Management of Data, pp. 151–162, May 2001

Kesseli, J., Ramo, P., Yli-Harja, O.: Inference of Boolean models of genetic networks using monotonic time transformations. Control, Communications and Signal Processing. In: First International Symposium on, pp. 759–762, 21–24 March 2004

Khan, J., et al.: Classification and diagnostic prediction of cancers using gene expression profiling and artificial neural networks. Nat. Med. **7**(6), 673–679 (2001)

Kimball, R.: The Data Warehouse Toolkit. Wiley, New York (1996)

Klein, S.: Knowledge visualization in practice: challenges for future corporate communication. In: Ninth International Conference on Information Visualisation (IV'05), London, England, 6–8 July 2005

Klir, G.: The many faces of uncertainty. In: Ayyub, B.M., Gupta, M.M. (eds.) Uncertainty Modeling and Analysis: Theory and Applications, pp. 3–19. Elsevier Science, USA (1994)

Kodogiannis, V., Lolis, A.: Forecasting financial time series using neural network and fuzzy system-based techniques. Neural Comput. Appl. **11**, 90–102 (2002)

Korenberg, M.: A robust orthogonal algorithm for system identification and time-series analysis. Biol. Cybern. **60**(4), 267–276 (1989)

Kosuke, I., King, G., Lau, O.: Toward a common framework for statistical analysis and development. J. Comput. Graph. Stat. **17**(4), 892–913 (2008)

Krishna, A., Narayanan, A., Keedwell, E.: Reverse engineering gene networks with artificial neural networks. In: Adaptive and Natural Computing Algorithms. Springer, Vienna (2005)

Kundaje, A., Middendorf, M., Gao, F., Wiggins, C., Leslie, C.: Combining sequence and time series expression data to learn transcriptional modules. IEEE/ACM Trans. Comput. Biol. Bioinform. **2**(3), 194–202 (2005)

Lahdesmaki, H., Shmulevich, I., Yli-Harja, O.: On learning gene regulatory networks under the Boolean network model. Mach. Learn. **52**, 147–167 (2003)

Laird, N., Ware, J.: Random effects models for longitudinal data. Biometrics **38**, 963–974 (1982). doi:10.2307/2529876

Lipinski, P., Winczura, K., Wojcik, J.: Building risk-optimal portfolio using evolutionary strategies. In: Applications of Evolutionary Computing: EvoWorkshops 2007, EvoCOMNET, EvoFIN. Springer (2007)

Lamers, S., et al.: Prediction of R5, X4, and R5X4 HIV-1 Coreceptor Usage with Evolved Neural Networks. IEEE/ACM Trans. Comput. Biol. Bioinform. **5**(2), 291–300 (2008)

Langmead, C., McClung, C., Donald, B.: A maximum entropy algorithm for rhythmic analysis of genome-wide expression patterns. In: Bioinformatics Conference 2002, IEEE, pp. 237–245 (2002)

Lapedes, A.S., Farber, R.: Non-linear signal processing using neural networks: prediction and system modeling. In: Technical Report LA-UR-87. Los Alamos National Laboratory (1987)

Law, R., Au, N.: A neural network model to forecast Japanese demand for travel to Hong Kong. Tourism Manage. **20**(1), 89–97 (1999)

Lee, W., Yang, K.: Applying intelligent computing techniques to modeling biological networks from expression data. Genomics, Proteomics Bioinformatics **6**(2), 111–120 (2008)

Leigh, W., Purvis, R., Ragusa, J.M.: Forecasting the NYSE composite index with technical analysis, pattern recognizer, neural network, and genetic algorithm: a case study in romantic decision support. Decis. Support Syst. **32**, 361–377 (2002)

Lenser, S., Veloso, M.: Non-parametric time series classification. 2005. In: Proceedings of the 2005 IEEE International Conference on Robotics and Automation, Barcelona, Spain, 18–22 April 2005

Lenstra, A.: Integer Factoring – Designs, Codes and Cryptography, vol. 19, pp. 101–128. Kluwer, Netherlands (2000)

Lewis, J.: Fast template matching. Vision Interface, 120–123 (1995)

Li, S., Wu, L., Zhang, Z.: Constructing biological networks through combined literature mining and microarray analysis: a LMMA approach. Bioinformatics **22**(17), 2143–2150 (2006a)

Li, X., Rao, S., Jiang, W., Li, C., Xiao, Y., Guo, Z., Zhang, Q., Wang, L., Du, L., Li, J., Li, L.: Discovery of time-delayed gene regulatory networks based on temporal gene expression profiling. BMC Bioinform. **7**, 26 (2006b). doi:10.1186/1471-2105-7-26

Liao, S., et al.: Code optimization techniques for embedded DSP microprocessors. In: Design Automation for Embedded Systems. Springer, Germany (1998)

Lin, J., et al.: Visually mining and monitoring massive time series. In: Proceedings of the Tenth ACM SIGKDD International Conference on Knowledge Discovery and Data Mining, pp. 460–469, Seattle, USA (2004)

Lin, J. et al.: A Symbolic representation of time series, with implications for streaming algorithms. In: Proceedings of the 8th ACM SIGMOD Workshop on Research Issues in Data Mining and Knowledge Discovery, pp. 2–11, San Diego, California, USA (2003)

Lipinski, P., Winczura, K., Wojcik, J.: Biilding risk-optional partfolio using evolutionary strategies. In: Applications of evolutionary computing: Evoworkshops 2007, EvoCOMNET, evoFIN. Springer (2007)

Liu, B., Cui, Q., Jiang, T., Ma, S.: A combinational feature selection and ensemble neural network method for classification of gene expression data. BMC Bioinform. **5**, 136 (2004). doi:10.1186/1471-2105-5-136

Liu, C.: Introduction to Combinatorial Mathematics. McGraw-Hill, New York (1968)

Liu, Y., Eyal, E., Bahar, I.: Analysis of correlated mutations in HIV-1 protease using spectral clustering. Bioinformatics **24**(10), 1243–1250 (2008)

Lynch, N., Shvartsman, A.: Communication and data sharing for dynamic distributed systems. In Future Directions in DC 2002. LNCS **2584**, 62–67 (2003)

Maoz, D., Sternberg, A., Leibowitz, E.: Astronomical Time Series. Kluwer, New York (1997)

Maqsood, I., Abraham, A.: Weather analysis using ensemble of connectionist learning paradigms. Appl. Soft Comput. **2007**(7), 995–1004 (2007)

Maqsood, I., Khan, M., Abraham, A.: An ensemble of neural networks for weather forecasting. Neural Comput. Appl. **13**, 112–122 (2004)

Maraziotis, I., Dragomir, A., Bezerianos, A.: Gene networks reconstruction and time-series prediction from microarray data using recurrent neural fuzzy networks. IET Syst. Biol. **1**(1), 41–50 (2007)

Markham, I.S., Ragsdale, C.T.: Combining neural networks and statistical predictions to solve the classification problem in discriminant analysis. Decision Sci. **26**(2), 229–241 (1995)

Martin, S., Zhang, Z., Martino, A., Faulon, J.: Boolean dynamics of genetic regulatory networks inferred from microarray time series data. Bioinformatics **23**(7), 866–874 (2007)

Masih, A.M., Masih, M.R.: Are Asian stock market fluctuations due mainly to intra-regional contagion effects? Evidence based on Asian emerging stock markets. Pac. Basin Financ. J. **7**(3–4), 251–282 (1999)

Masih, A.M., Masih, M.R.: Long and short-term dynamic causal transmission amongst international stock markets. J. Int. Money Financ. **20**(4), 563–587 (2001)

Mattera D., Haykin S.: Support vector machines for dynamic reconstruction of a chaotic system. In: Schölkopf, B., Burges, C.J.C., Smola, A.J. (eds.) Advances in Kernel Methods – Support Vector Learning, pp. 211–242. MIT, Cambridge, MA (1999)

McCulloch, W.W., Pitts, W.: A logical calculus of the ideas imminent in nervous activity. Bull. Math. Biophys. **5**, 115–133 (1943)

McGarry, K., Wermter, S., Maclntyre, J.: Hybrid neural systems: from simple coupling to fully integrated neural networks. Neural Comput. Surv. **2**, 62–93 (1999)

McLachlan, G.: Discrimination Analysis and Statistical Pattern Recognition. Wiley Interscience, USA (2004)

McQuarrie, A., Tsai, C.: Regression and Time Series Model Selection. World Scientific, Singapore (1998)

Menard, M.: Fuzzy clustering and switching regression models using ambiguity and distance rejects. Fuzzy Set. Syst. **133**, 363–399 (2001)

Meusinger, H., Scholz, R.-D., Irwin, M., Lehmann, H.: QSOs from the variability and proper motion survey in the M3 field. A&A **392**, 851 (2002)

Molina, J., et al.: Segmentation and classification of time-series: real case studies. In Intelligent data engineering and automated learning – IDEAL 2009, pp. 743–750. Springer, Germany (2009)

Moore, J., Boczko, E., Summar, M.: Connecting the dots between genes, biochemistry, and disease susceptibility: systems biology modeling in human genetics. Mol. Genet. Metab. **84**(2), 104–111 (2005)

Morgan, R.E., Hunt, S.D.: Determining marketing strategy – a cybernetic systems approach to scenario planning. Eur. J. Marketing **36**(4), 450–478 (2002)

Mukhopadhyay, N., Chatterjee, S.: Causality and pathway search in microarray time series experiment. Bioinformatics **23**(4), 442–449 (2007)

Muller K., Smola A., Ratsch G., Scholkopf B., Kohlmorgen J., Vapnik V.: Predicting time series with support vector machines. In: Artificial Neural Networks ICANN'97, Lecture Notes in Computer Science, vol. 1327, pp. 999–1004. Springer (1997)

Neagu, D., Palade, V.: A Neuro-fuzzy approach for functional genomics data interpretation and analysis. Neural Comput. Appl. **12**, 153–159 (2003)

Newberg, H., Yanny, B.: ApJS **113**, 89 (1997)

Nemati, H., et al.: Knowledge warehouse: an architectural integration of knowledge management, decision support, artificial intelligence and data warehousing. Decis. Support Syst. **33**(2), 143–161 (2002)

Ng, M., Wong, J.: Clustering categorical data sets using tabu search techniques. Pattern Recogn. **35**, 2783–2790 (2002)

Nikkilä, J., et al.: Analysis and visualization of gene expression data using self-organizing maps. Neural Networks **15**(8–9), 953–966 (2002)

Olds, J., Steadman, K.: Cross-platform computational techniques for analysis code integration and optimization. In: 7th AIAA/USAF/NASA/ISSMO Symposium on Multidisciplinary Analysis and Optimization. St. Louis, MO, USA, 2–4 September 1998

Oliva, A., Rosebrock, A., Ferrezuelo, F., Pyne, S., Chen, H., Skiena, S., Futcher, B., Leatherwood, J.: The cell cycle-regulated genes of Schizosaccharomyces pombe. PLoS Biol. **3**(7), 1239–1260 (2005)

Oliveira, S., Seok, S.: A matrix-based multilevel approach to identify functional protein modules. Int. J. Bioinform. Res. Appl. **4**(1), 11–27 (2008)

Olson, G., Zimmerman, A., Bos, N. (eds.): Scientific Collaboration on the Internet. MIT, USA (2008)

Opitz, D., Maclin, R.: Popular ensemble methods: an empirical study. J. Artif. Intell. Res. **11**, 169–198 (1999)

Ostrom, J.: Time Series Regression. Sage, Beverly Hills, CA, USA (1990)

O'Sullivan, E., Rassel, G.: Research Methods for Public Administrators, 3rd edn. Longman, UK (1999)

Pai, P.F., Hong, W.C.: An improved neural network model in forecasting arrivals. Ann. Tourism Res. **32**(4), 1138–1141 (2005)

Palmer, A., Montano, J.J., Sese, A.: Designing an artificial neural network for forecasting tourism time series. Tourism Manage. **27**(5), 781–790 (2006)

Parsopoulos, K., Vrahatis, M.: Recent approaches to global optimization problems through particle swarm optimization. Nat. Comput. **1**(2–3), 235–306 (2002)

Peck, A.: The role of graphic art in modern scientific communication. In: Proceedings of Society for Technical Communication, 1995. http://www.stc.org/confproceed/1995/PDFs/PG425426.PDF

Percival, D., Walden, A.: Spectral Analysis for Physical Applications. Cambridge University Press, Cambridge, UK (1993)

Perlman, E., Java, A.: Predictive mining of time series data in astronomy. Astron. Data Anal. Softw. Syst. XII ASP Conf. Ser. **295**, 431 (2003)

Pinna, G., Maestri, R., Cesare, A.: Application of time series spectral analysis theory: analysis of cardiovascular variability signals. J. Med. Biol. Eng. Comput. **34**(2), 142–148 (1996)

Pittenger, A.: An Introduction to Quantum Computing Algorithms. Springer, Germany (2000)

Priestley, M.: Spectral Analysis and Time Series. Academic, London, UK (1982)

Principe, J., Euliano, N., Lefebvre, W.: Neural and Adaptive Systems: Fundamentals Through Simulations. Wiley, USA (2000)

Prinzie, A., Van den Poel, D.: Incorporating sequential information into traditional classification models by using an element/position-sensitive SAM. Decis. Support Syst. **42**(2), 508–526 (2006)

Rafiei, D., Mendelzon, A.: Efficient retrieval of similar time sequence using DFT. In: The 5th International Conference on Foundations of Data Organization, Kobe, Japan, November 1998.

Rajan, A., Rawat, A., Verma, R.: Virtual computing grid using resource pooling. In: ICIT 2008 International Conference on Information Technology, pp. 59–64, 2008

Ratanamahatana, C., Keogh, E.: Three myths about dynamic time warping data mining. In: Proceedings of the 5th SIAM International Conference on Data Mining, pp 506–510, 2005

Refenes, A.N., Zapranis, A., Francis, G.: Stock performance modeling using neural networks: a comparative study with regression models. Neural Networks **5**, 961–970 (1994)

Refinetti, R.: Non-stationary time series and the robustness of circadian rhythms. J. Theor. Biol. **227**(4), 571–581 (2003)

Reuter, A.: Methods for parallel execution of complex database queries. Parallel Comput. **25**(13–14), 2177–2188 (1999)

Richards, G., et al.: Spectroscopic Target Selection in the Sloan Digital Sky Survey: The Quasar Sample. Astron. J. **123**, 2945 (2002)

Richards, G., et al.: Efficient photometric selection of quasars from the sloan digital sky survey: 100,000 $z < 3$ quasars from data release one. ApJS **155**, 257 (2004)

Richards, G., et al.: Efficient photometric selection of quasars from the sloan digital sky survey: II. ~ 1,000,000 quasars from data release six. ApJS **180**, 67 (2009)

Richman, J., Moorman, J.: Physiological time-series analysis using approximate entropy and sample entropy. Am. J. Physiol. – Heart C **278**, 2039–2049 (2000)

Ritchie, M., et al.: Exploring epistasis in candidate genes for rheumatoid arthritis. BMC Proc. **1**(Suppl. 1), S70 (2007)

Ronen, M., Shabtai, Y., Guterman, H.: Rapid process modelling-model building methodology combining supervised fuzzy-clustering and supervised neural networks. Comput. Chem. Eng. **22**, S1005–1008 (1998)

Rosenblatt, F.: The perceptron: a probabilistic model for information storage and organization in the brain. Psycho. Rev. **65**(6), 386–408 (1958)

Rowland, J.: Model selection methodology in supervised learning with evolutionary computation. Biosystems **72**(1–2), 187–196 (2003)

Saeed, M., Mark, R.: A novel method for the efficient retrieval of similar multiparameter physiologic time series using wavelet-based symbolic representations. AMIA Annu. Symp. Proc. 679–683 (2006)

Sakamoto, E., Iba, H.: Inferring a system of differential equations for a gene regulatory network by using genetic programming. In: IEEE Proceedings of the Congress on Evolutionary Computation, vol. 1, pp. 720–726, May 2001

Sandage, A., Wyndham, J.D.: ApJ **141**, 328 (1965)
Sarkar, M., Yegnanarayana, B., Khemani, D.: Backpropagation learning algorithms for classification with fuzzy mean square error. Pattern Recognit. Lett. **19**, 43–51 (1998)
Schmidtke, P.C., Cowley, A., Crane, J., Taylor, V., McGrath, T., Hutchings, J., Crampton, D.: AJ **117**, 927 (1999)
Schneider, D.P., et al.: AJ **134**, 102 (2007). arXiv:0704.0806
Schneider, D.P., et al.: The sloan digital sky survey quasar catalog. II. first data release. Astron. J. **126**(6), 2579–2593 (2003)
Scholz, R., Meusinger, H., Irwin, M.: A UBV/variability/proper motion QSO survey from Schmidt plates I. Method and success rate. A&A **325**, 457 (1997)
Schuster, A.: On the investigation of hidden periodicities with application to a supposed 26 day period of meteorological phenomena. Terr. Magn. Atmos. Electr. **3**, 13–41 (1898)
Schwarzenberg-Czerny, A.: MNRAS **241**, 153 (1989)
Shaw, S., DeFigueiredo, R.: Structural processing of waveforms as trees. IEEE Trans. Acoust. Speech. **38**, 2 (1990)
Shi, Y., Mitchell, T., Bar-Joseph, Z.: Inferring pairwise regulatory relationships from multiple time series datasets. Bioinformatics **23**(6), 755–763 (2007)
Shin, M., Sekora, M., Byun, Y.: Detecting variability in massive astronomical time-series data I: application of an infinite Gaussian mixture model. Accepted for publication in Monthly Notices of the Royal Astronomical Society **400**(4), 1897–1910 (2009)
Shmulevich, I., Dougherty, E., Kim, S., Zhang, W.: Probabilistics Boolean networks: a rule-based uncertainty model for gene regulatory networks. Bioinformatics **18**(2), 261–274 (2002)
Shumway, R., Stoffer, D.: Time Series Analysis and Its Applications, 2nd edn. Springer, Germany (2006)
Sims, C.: Macroeconomics and reality. Econometrica **48**, 1–48 (1980)
Singhal, A., Seborg, D.: Clustering multivariate time-series data. J. Chemometr. **19**, 427–438 (2005)
Smith, K.A., Gupta, J.N.D.: Neural networks in business: techniques and applications for the operations researcher. Comput. Oper. Res. **27**, 1023–1044 (2000)
Smola, A., Scholkopf, B.: A tutorial on support vector regression. Stat. Comput. **14**, 199–222 (2004)
Soydemir, G.: The impact of the movements in US three month treasury bill yields on the equity markets in Latin America. Appl. Financ. Econ. **12**(2), 77–84 (2002)
Spangl, B., Dutter, R.: Estimating spectral density functions robustly. REVSTAT – Stat. J. **5**(**1**), 41–61 (2007)
Spath, H.: Cluster Analysis Algorithms. Ellis Horwood, Chichester, UK (1980)
Sumi, T., et al.: Variability-selected QSO candidates in OGLE-II Galactic bulge fields. MNRAS **357**, 331 (2005)
Swanson, N.R., White, H.: A model selection approach to real-time macroeconomic forecasting using linear models and artificial neural networks. Rev. Econ. Stat. **79**(4), 540–550 (1997)
Syeda-Mahmood, T.: Clustering time-varying gene expression profiles using scale-space signals. Bioinformatics Conference, 2003. CSB 2003. In: Proceedings of the 2003 IEEE, pp. 48 – 56, 11–14 August 2003
Tabus, I., Astola, J.: Clustering the non-uniformly sampled time series of gene expression data. Signal Processing and Its Applications. In: Proceedings of the Seventh International Symposium on, Vol. 2, pp. 61–64, 1–4 July 2003
Tabus, I., Giurcaneanu, C., Astola, J.: Genetic networks inferred from time series of gene expression data. Control, Communications and Signal Processing. In: First International Symposium on, pp. 755–758, 21–24 March 2004
Tao, C.:. Robust control of systems with fuzzy representation of uncertainties. Soft Comput. – A Fusion Found. Methodol. Appl. **8**(3), 163–172 (2004)
Tao, Y., et al.: Exploiting similarity of subqueries for complex query optimization. In: Database and Expert Systems Applications. Springer, Berlin (2003)
Taylor, J., et al.: Application of metabolomics to plant genotype discrimination using statistics and machine learning. Bioinformatics **18**(Suppl. 2), 241–248 (2002)

Tennenhouse, D., Wetherall, D.: Towards an active network architecture. ACM SIGCOMM Comput. Commun. Rev. (2007)

Tessier, R., Burleson, W.: Reconfigurable computing for digital signal processing: a survey. J. VLSI Signal Proc. **28**, 7–27 (2001)

Tettamanzi, A., Tomassini, M.: Soft Computing, Integrating Evolutionary, Neural, and Fuzzy Systems. Springer, Germany (2001)

Theodoridis, S., Koutroumbas, K.: Pattern Recognition, 2nd edn. Academic, 2003

Tickle, A., et al.: The truth will come to light: directions and challenges inextracting the knowledge embedded within trained artificial neuralnetworks. IEEE Trans. Neural Networ. **9**(6), 1057–1068 (1998)

Tu, J.V.: Advantages and disadvantages of using artificial neural networks versus logistic regression for predicting medical outcomes. J. Clin. Epidemiol. **49**(11), 1225–1231 (1996)

Tudela, R., et al.: Full complex Fresnel holograms displayed on liquid crystal devices. J. Opt. A: Pure Appl. Opt. **5**, s189–s194 (2003)

Tufte, E.: Envisioning Information. Graphics Press, Cheshire, CT, USA (1990)

Tufte, E.: The Visual Display of Quantitative Information. Graphics Press, USA (1992)

Tytell, E., Standen, E., Lauder, G.: Escaping Flatland: three-dimensional kinematics and hydrodynamics of median fins in fishes. J. Exp. Biol. **211**, 187–195 (2008)

Udalski, A., Szymanski, M., Kubiak, M., Pietrzynski, G., Soszynski, I., Wozniak, P., Zebrun, K., Szewczyk, O., Wyrzykowski, L.: The optical gravitational lensing experiment. BVI maps of dense stellar regions. III. The Galactic Bulge. Acta Astron. **52**, 217–240 (2002)

Uysal, M., Roubi, M.S.E.: Artificial neural networks versus multiple regression in tourism demand analysis. J. Travel Res. **38**, 111–118 (1999)

Valdes, P., et al.: Frequency domain models of the EEG. Brain Topogr. **4**(4), 309–319 (1992)

Vandersypen, L., et al.: Separability of very noisy mixed states and implications for NMR quantum computing. Phys. Rev. Lett. **83**, 1054–1057 (1999)

Varanini, M., et al.: Adaptive modelling of biological time series for artifact detection. Computers in Cardiology, Proceedings, pp. 695–698 (1993)

Vellido, A., Lisboa, P.J.G., Vaughan, J.: Neural networks in business: a survey of applications (1992–1998). Expert Systems Appl. **17**, 51–70 (1999)

Vestrand, W., Theiler, J., Wozniak, P.: Unsolved problems in observational astronomy. II. Focus on rapid response – mining the sky with "thinking" telescopes. Astronomische Nachrichten **325**(6–8), 477–482 (2004)

Voss, A., et al.: The application of methods of non-linear dynamics for the improved and predictive recognition of patients threatened by sudden cardiac death. Cardiovasc. Res. **31**, 419–433 (1996)

Vries, W., et al.: Structure function analysis of long-term quasar variability. Astron. J. **129**, 615 (2005)

Wagner, N., Michalewicz, Z.: An analysis of adaptive windowing for time series forecasting in dynamic environments: further tests of the DyFor GP model. Genetic And Evolutionary Computation Conference. In: Proceedings of the 10th Annual Conference on Genetic and Evolutionary Computation. Atlanta, GA, USA (2008)

Wang, W., et al.: A novel hybrid intelligent model for financial time series forecasting and its application. In: International Conference on Business Intelligence and Financial Engineering. IEEE (2009)

Wang, X., Li, A., Jiang, Z., Feng, H.: Missing value estimation for DNA microarray gene expression data by Support Vector Regression imputation and orthogonal coding scheme. BMC Bioinform. **7**, 32 (2006a)

Wang, X., Smith, K., Hyndman, R.: Characteristic-based clustering for time series data. Data Min. Knowl. Disc. **13**(3), 335–364 (2006b)

Wang, X., et al.: Short time-series microarray analysis: methods and challenges. BMC Syst. Biol. **2**, 58 (2008). doi:10.1186/1752-0509-2-58

Weaver, B.: Spectral classification of unresolved binary stars with artificial neural networks. Astrophys. J. **541**, 298–305 (2000)

Weisstein, E.: Discrete Fourier Transform. MathWorld – A Wolfram Web Resource, 2009. http://mathworld.wolfram.com/DiscreteFourierTransform.html.

Welch, D., Stetson, P.: Robust variable star detection techniques suitable for automated searchs: new results for NGC 1866. AJ **105**, 1813–1821 (1993)

Werbos, P.J.: Beyond regression: new tools for prediction and analysis in the behavioral sciences. Ph.D. thesis, Harvard University (1974)

White, H.: Learning in artificial neural networks: a statistical perspective. Neural Comput. **1**, 425–464 (1989)

Widrow, B.: Generalization and information storage in networks of adaline neurons. Self-Organizing Systems, pp. 435–461. Spartan Books, Washington, DC (1959)

Wiener, N.: The theory of prediction. In: The Theory of Prediction. McGraw-Hill, New York, USA (1956)

Williamson, A.G.: Refining a neural network credit application vetting system with a genetic algorithm. J. Microcomput. Appl. **18**, 261–277 (1995)

Winter, S., et al.: Overcomplete BSS for convolutive mixtures based on hierarchical clustering. In: Independent Component Analysis and Blind Signal Separation. Springer, Berlin, Germany (2004)

Wischik, D., Handley, M., Braun, M.: The resource pooling principle. ACM SIGCOMM Comput. Commun. Rev. **38**(5), 47–52 (2008)

Witt, S.F., Witt, C.A.: Forecasting tourism demand: A review of empirical research. Int. J. Forecasting **11**(3), 447–475 (1995)

Wolf, C., Meisenheimer, K., Roser, H.: Object classification in astronomical multi-color surveys. Astron. Astrophys. **365**, 660–680 (2001)

Wo'zniak, P.R., Udalski, A., Szyma'nski, M., Kubiak, M., Pietrzy'nski, G., Soszy'nski, I., Zebru'n, K.: Acta Astron. **52,** 129 (2002)

Wu, F., Zhang, W., Kusalik, A.: Determination of the minimum sample size in microarray experiments to cluster genes using k-means clustering. Bioinformatics and Bioengineering, 2003. In: Proceedings of the Third IEEE Symposium on, pp. 401–406, 10–12 March 2003

Wu, H., Siegel, M., Khosla, P.: Vehicle sound signature recognition by frequency vector principal Component Analysis. In: IEEE Instrumentation and Measurement Technology Conference, St. Paul, MN, USA, 18–20 May 1998

Wu, X., Fung, M., Flitman, A.: Forecasting stock market performance using hybrid intelligent system. In: Computational Science – ICCS, Springer, Berlin (2001)

Xu, J., et al.: A new nonlinear similarity measure for multichannel biological signals. In: Proceedings of International Joint Conference on Neural Networks, Orlando, FL, USA, 12–17 August 2007

Yang, Y., Suen, J., Brynildsen, M., Galbraith, S., Liao, J.: Inferring yeast cell cycle regulators and interactions using transcription factor activities. BMC Genomics **6**, 90, 2005. doi:10.1186/1471-2164-6-90

Yankov, D., Keogh, E., Rebbapragada, U.: Disk aware discord discovery: finding unusual time series in terabyte sized datasets. Knowl. Inform. Syst. **17**(2), 241–262 (2008)

Yao, J.T., Tan, C.L.: A case study on using neural networks to perform technical forecasting of Forex. Neurocomputing **34**, 79–98 (2000)

Yeang, C., Jaakkola, T.: Time series analysis of gene expression and location data. Bioinformatics and Bioengineering, 2003. In: Proceedings of the Third IEEE Symposium on, pp. 305–312, 10–12 March 2003

Yeung, K., Ruzzo, W.: Principal component analysis for clustering gene expression data. Bioinformatics **17**(9), 763–774 (2001)

Yoshioka, T., Ishii, S.: Clustering for time-series gene expression data using mixture of constrained PCAS. Neural Inf. Proces., ICONIP '02, **v5**, 2239–2243 (2002)

Yu, L., Wang, S.Y., Lai, K.K.: A novel nonlinear ensemble forecasting model incorporating GLAR and ANN for foreign exchange rate. Comput. Oper. Res. **32**(10), 2523–2541 (2005)

Zadeh, L., et al.: Fuzzy Sets, Fuzzy Logic. World Scientific Press, Fuzzy Systems (1996)

Zhang, G. (ed.): Neural Networks in Business Forecasting. IGI Publishing (2004)

Zhang, G.P.: Time series forecasting using a hybrid ARIMA and neural network model. Neurocomputing **50**, 159–175 (2003)

Zhang, L., Zhang, A., Ramanathan, M.: Fourier harmonic approach for visualizing temporal patterns of gene expression data. Bioinformatics Conference, 2003. CSB 2003. In: Proceedings of the 2003 IEEE, pp. 137–147, 11–14 August 2003

Zhang, W., Bai, C., Liu, G.: Neural network modeling of ecosystems: a case study on cabbage growth system. Ecol. Model. **201**, 317–325 (2007)

Zhang, Y., Zhao, Y.: BBN, ADTree and MLP comparison in separating quasars from large survey catalogues. Chin. J. Astron. Astrophys. **15**, 10 (2008)

Zou, M., Conzen, S.: A new dynamic Bayesian network (DBN) approach for identifying gene regulatory networks from time course microarray data. Bioinformatics **21**(1), 71–79 (2005)

MIX
Papier aus verantwortungsvollen Quellen
Paper from responsible sources
FSC® C105338

If you have any concerns about our products,
you can contact us on
ProductSafety@springernature.com

In case Publisher is established outside the EU,
the EU authorized representative is:
Springer Nature Customer Service Center GmbH
Europaplatz 3, 69115 Heidelberg, Germany

Printed by Libri Plureos GmbH
in Hamburg, Germany